Studying Engineering

A Road Map to a Rewarding Career

Third Edition

by
Raymond B. Landis, Dean Emeritus
College of Engineering, Computer Science, and Technology
California State University, Los Angeles

Published by:
Discovery Press
Los Angeles, California
www.discovery-press.com

Permissions and Copyrights

Cover design by Dave McNutt

Chapter title illustrations by Brian Jefcoat

Case study of Nuna 3 solar car project by permission of Dutch energy company Nuon and Jorrit Lousberg, Student Team Leader, Delft University of Technology

Franklin Chang-Diaz photo in Chapter 1 courtesy of National Aeronautics and Space Administration

Studying Engineering: A Road Map to a Rewarding Career, Third Edition

Discovery Press/2007

10 9 8 7 6

ISBN 978-0-9646969-2-1

Inquiries and comments should be addressed to:

Raymond B. Landis, Ph.D.
Dean Emeritus of Engineering, Computer Science, and Technology
California State University, Los Angeles
Los Angeles, California 90032
E-mail: rlandis@calstatela.edu

Distributed by:

Legal Books Distributing
4247 Whiteside Street
Los Angeles, CA 90063
Website: www.legalbooksdistributing.com
Telephone: (323) 526-7110
(800) 200-7110 (From outside Los Angeles County)
E-mail: info@legalbooksdistributing.com

Books may be ordered by e-mail, by telephone, or on-line

TO KATHY

FOREWORD

When I was a sophomore in high school, I decided that I wanted to be a chemical engineer when I grew up. I could invent all sorts of reasons for this decision that would make me sound like an unusually wise and thoughtful 15-year-old, but they would all be lies.

The truth is at the time there was a great job market for engineers, and stories of red carpets and multiple job offers and outlandishly high starting salaries were laid on us regularly by teachers and counselors—and in my case, by parents. Just about every boy who could get B or better in math and science courses decided that he was born to be an engineer, and I saw no reason to buck the trend. Why *chemical* engineering? Because—sadly, this is also the truth—I had gotten a chemistry set for my birthday, and I thought pouring one liquid into another and having it turn green was seriously cool.

Like most of my engineering-bound classmates, I knew nothing about what engineers actually did for a living, and when I enrolled in chemical engineering at the City College of New York two years later I still knew nothing. There was a freshman orientation course, but it was just the old "Sleep 101" parade of unenthusiastic professors delivering dreary 40-minute sermons about civil engineering, mechanical engineering, and so forth. It's a wonder that this course didn't drive more people away from engineering than it informed and motivated. Perhaps it did.

My ignorance persisted for pretty much the next three years as I worked through the math and physics and chemistry and thermo and transport and circuits and all those other things you have to know to graduate in engineering, but only represent a small fraction of what engineers actually do. It wasn't until I got into the unit operations lab in my fourth year and then spent a summer in industry that I started to get a clue about what engineering is really about—figuring out why things aren't working the way they're supposed to and fixing them, and designing and building other things that work better or work just as well and cost less.

And what engineers did for a living was only the tip of the iceberg of what I didn't know as a freshman. In high school I rarely cracked a textbook and still came out with nearly straight A's, but it took only one college physics exam to let me know that the game had changed. I also left high school thinking I was a great writer, but the D+ on my first college English paper set me straight about that too. Plus, I didn't know how to take notes, summarize long reading assignments, prepare for and take tests, strike a good balance between school and the rest of my life. I could go on but you get the idea.

I eventually figured it all out, of course. If I hadn't, I wouldn't have graduated and gone on to be an engineering professor and the author of this foreword. Unfortunately, many of my classmates never did get it, and most of them were gone by the end of the second year. And I know they had the ability to succeed.

I don't think engineering school should be an academic obstacle course designed to weed out students who have the ability to succeed but lack basic study skills. If something is important for students to know, there's nothing wrong with giving them some guidance in figuring it out. We do that routinely with math and science and control and design. Why not do it with studying and learning?

That's where Ray Landis and *Studying Engineering* come in. The book is a compendium of everything I wish someone had told me in my freshman year. If I could have read it then, even if I had only absorbed a fraction of the wisdom it contains, I would have been spared the major headache of having to learn it the hard way. And if the book had been used in a first-year engineering course taught by a knowledgeable and supportive instructor, the next four years of my life would have been far less stressful, and many of my talented classmates who dropped out as freshmen and sophomores would instead have graduated with me.

Almost everything students need to know to succeed in engineering school is in *Studying Engineering*. Using a conversational tone and numerous real-world examples and anecdotes, Professor Landis paints a vivid picture of the vast range of things engineers do, the world-changing things some of them have done in the past, and the challenges to ingenuity and creativity that they routinely face. He also introduces students to the learning process—how it works, when and why it goes wrong, and how to avoid the pitfalls that have ensnared generations of engineering students including those unfortunate classmates of mine.

Moreover, *Studying Engineering* introduces its readers to themselves and to one another, providing insights into different ways people approach learning tasks and respond to instruction. Students who take this material to heart will gain a better understanding of their own strengths and weaknesses and will learn ways to capitalize on the former and overcome the latter. Their new knowledge will also improve their ability to communicate with their classmates and teammates. These insights and skills will serve them well throughout college and in their subsequent professional careers, whether or not they remain in engineering.

If you are an engineering educator who teaches first-year students, I invite you to think about the things you wish someone had told you when you were a freshman, and then use *Studying Engineering* to help convey those messages. If you are a student, I encourage you to pay attention to the book, because it's telling you things that are important. If you're going to succeed in engineering school, you'll need to learn those things, sooner or later. My advice is, make it sooner.

Richard M. Felder
Hoechst Celanese Professor Emeritus of Chemical Engineering
North Carolina State University
Raleigh, North Carolina

TABLE OF CONTENTS

PAGE

PREFACE
(Excerpted from First Edition of *Studying Engineering*)

[Note: I am taking the liberty of reprinting part of the preface from the first edition of this book I'm not sure whether people read prefaces, but I do hope you'll read this one. It does about as good a job as I could do to explain the philosophy that underlies this book. - R.B.Landis]

We aren't born knowing how to be effective. We learn how. We learn from our parents or guardians, from our teachers, from our peers, and from supervisors and mentors. We learn from workshops and seminars, from books, and from trial and error. Developing our effectiveness is a life-long process. Sometimes we get more help than other times. For example, when we join an organization as a professional, we generally receive lots of help. The organization benefits if we are successful, and so it takes steps to ensure that we are.

Industry executives are well aware that new engineering graduates have a long way to go before they can "earn their salary." New engineering hires are thus provided with formal training, on-the-job training, close supervision, progressively more challenging assignments, rotating work assignments, and time to mature.

Strangely, when new students (or, in fact, new faculty) come to the university, they are left primarily on their own to figure out how to be successful. Academic organizations seem more interested in evaluating their newest members than in doing things to ensure that they succeed.

Within engineering education, this "sink or swim" approach is not working. Only about 40 percent of students who start engineering study ever graduate. Most drop out, flunk out, or change their majors. And many of those who do graduate fail to work up to their full potential.

Even deans of engineering need training. As a new dean, I had four separate consultants in for two days each to teach me (and my school's faculty) how to be effective in preparing for our upcoming accreditation process. In addition, I have participated in formal training in personnel management, fund raising, Total Quality Management, computer technology, and teaching methods.

If new engineering graduates and new engineering deans need orientation, training, mentoring, and time to mature to be effective, how is it that as engineering educators we expect our students to know how to go about the task of engineering study the day they arrive?

Sometimes it appears that we don't want our students to succeed. We seem to go out of our way to avoid helping our students learn to be effective. Our view of subjects like professional development, academic success strategies, personal development, and orientation is that they are not "academic." We are reluctant to find room for them in our already full curricula.

But it even goes farther than that. We sometimes seem pleased by the fact that many of our students don't succeed. We find satisfaction in the view that "not everyone can be an engineer." Our approach is to put up a difficult challenge and believe that we have done a service to the profession by "weeding out" those who don't measure up. We tend to hold the black-and-white view that "some have it, and some don't."

If it were true that some students have it and some don't, then it probably wouldn't make sense to devote time and effort to helping students develop the skills they need to succeed. It wouldn't make a difference anyway. But this is one heck of a view for educators to have.

The good news, however, is that engineering education in the United States appears to be undergoing a revolution. We are in the process of a shift from the "sink or swim" paradigm to one of "student development." Engineering colleges across the nation are revising their freshman-year curricula with the primary goal of enhancing student success.

Although much of this curricular change involves moving more engineering content in areas such as design, graphics, computing, problem solving, and creativity into the freshman year, I hope that many engineering programs will find room for the "student development" content of this book in their freshman-year curriculum.

The basic premise of this book is that a small amount of time spent working with students on how to be effective early on can have an enormous payoff through the remainder of their college experience.

Raymond B. Landis
June, 1995

PREFACE TO THE THIRD EDITION

Studying Engineering, Third Edition has been updated and expanded. The primary change is that the key chapter—Chapter 3, "Academic Success Strategies"—has been expanded substantially and divided into three chapters. The new Chapters 3, 4, and 5 provide significant new information to aid students in taking advantage of both the teaching part and the learning part of the *teaching/learning* process. All of the dated material throughout the book has been updated, and a wealth of useful new Internet sites has been added. Finally, reflection exercises have been included as a new feature to assist students in processing the material.

Chapter 1 lays the foundation for the entire book. The process of achieving success in engineering study is introduced. Key elements of the success process—goal identification, goal clarification, and behavioral and attitudinal change—are presented. Three models that will help students understand what is meant by a quality education and how to go about getting that education are also introduced. The important topic of "Structuring Your Life Situation" has been moved into this chapter to ensure that students get exposed to it early on.

Chapter 2 addresses the subject of professional development. One of the primary purposes of the chapter is to motivate students through an increased understanding of the engineering profession and an increased awareness of the rewards and opportunities that will come to them if they are successful in graduating in engineering.

Chapter 3 provides an overview of the *teaching/learning* process. Various types of learning—cognitive; psychomotor; and affective—are described. Preferred learning styles and teaching styles are presented and contrasted. General guidelines for improving the learning process are provided, and a summary of the most common mistakes students make is presented along with approaches and strategies for avoiding these mistakes.

Chapter 4 provides guidance on how to get the most out of the teaching process. The importance of getting off to a good start is emphasized. Strategies for taking full advantage of lectures—including listening skills, note-taking skills, and questioning skills—are presented and discussed. Approaches for making effective use of professors are described in detail.

Chapter 5 guides students in designing their learning process. Two important skills for learning—reading for comprehension; and analytical problem solving—are covered. Approaches for organizing the learning process, including important time management skills, are discussed. Study skills that are relevant to math/science/engineering coursework are presented. Approaches for making effective use of peers through collaborative learning and group study are described in detail.

Chapter 6 focuses on the important subject of personal growth and development. A *Student Success Model* is presented to assist students in understanding the process of making behavioral and attitudinal changes essential to success in engineering study. Important personal development topics—understanding self, appreciating differences, personal assessment, communication skills, and health and wellness—are included in this chapter.

Chapter 7 addresses four extracurricular activities that can greatly enhance the quality of a student's education: (1) student organizations; (2) engineering projects; (3) pre-professional employment; and (4) service to the university.

Chapter 8 provides an orientation to the engineering education system including faculty, curriculum, students, facilities, administration, and institutional commitment. Academic regulations, student ethics, and opportunities for graduate education are also covered in this chapter.

The target audience for the book is first-year engineering students; therefore it is ideally suited for use in an *Introduction to Engineering* course that has a "student development/student success" objective. Much of what is in the book has direct application to the community college experience, and the topics that are specific to the four-year university experience can provide community college students with a preview of what they will encounter when they transfer to four-year institutions.

High school students considering engineering as their college major will find the book useful as well. Engineering faculty can turn to it as a resource for ideas they can convey to students in formal and informal advising sessions or in the classroom. Deans of engineering have indicated that the book contains material to help them prepare talks they give to high school students and first-year engineering students.

This book was the outgrowth of more than 30 years of teaching *Introduction to Engineering* courses. Much of the material was developed through brainstorming exercises with students. My greatest thanks go to the many students who contributed to the evolution of the ideas in this book. Thanks also go to the many engineering professors who have used the book since the *First Edition* was published in 1995. Those who provided valuable feedback on the *Second Edition* include: Hillar Unt, Sam Landsberger, Keith Sinkhorn, Janet Meyer, Jim Thomas, Dan Gulino, Wes Grebski, Ali Kujoory, Ken Brannan, Herb Schroeder, David Jackson, Blair Rowley, Ron Musiak, Brinda Subramaniam, Rick Dalrymple, Karen Groppi, Jim Lallas, Jose Ramirez, Jim Butler, Medica Denton, Marty Sirowatka, Hani Saad, Dick Otte, Mike Kelly, Dave Lueders, Dan Justice, Noel Caldwell, Zahir Khan, Gary Mishra, Daniel Styer, Laura Demsetz, Susan Sherod, Nick Arnold, Vince Bertsch, Chris Lenz, and Michael Read.

Many people contributed directly or indirectly to the creation of the book—both its original and its revised form. Much credit goes to my partner Martin Roden for encouraging me to self-publish the book and for his invaluable role in making it a reality. Great thanks to Dave McNutt for putting his extraordinary artistic talent into creating such a wonderful cover design. I also want to express my appreciation to William Gehr, President of Legal Books Distributing, and to his staff—particularly Mike O'Mahony, Ted Rogers, and Ruben Galindo—for handling the distribution of the book so capably and with so much care and concern.

I would also like to thank the following individuals who provided detailed feedback on various drafts of the *Third Edition*: Richard Felder, Jeff Froyd, Marty Roden, Darrell Guillaume, Eric Soulsby, Laurie Woollacott, Sarah Lifton, and Dom Dal Bello. Thanks also to Mark Regets and Nirmala Kannankutty at NSF and Michael Gibbons at ASEE for their help with key engineering education and employment data.

I would like to particularly acknowledge my wife Kathy Landis, who wrote the excellent section on Communication Skills in Chapter 6 and who did major editing and rewriting of the first two editions of the book. Her gifts as a writer and editor have made the book much easier to read and understand.

Raymond B. Landis
February, 2007

CHAPTER 1
Keys To Success In Engineering Study

Success is getting what you want;
happiness is wanting what you get.

Dale Carnegie

INTRODUCTION

This chapter introduces you to engineering study—both the process that will ensure you succeed, and the benefits you will get from doing so.

First, we make our best effort to convince you that you can do it: that success in engineering study, like success in anything you attempt, is a process that you can learn and master just as the many, many other successful students who came before you did.

We point out, however, a mindset that keeps some high ability, well-prepared students from mastering that process—overconfidence. Students who naively assume that their ability will carry them through engineering study as it did in high school can have a "rude awakening."

Next, we discuss two concepts fundamental to success—"goal identification" and "goal clarification." We also emphasize the importance of strongly committing to your goals once you have identified and clarified them.

Then, we present three important keys to success in engineering study:

Effort - *Work hard*
Approach - *Work smart*
Attitude - *Think positively*

As these keys to success reflect, achieving any challenging goal depends largely on your attitudes and behaviors—and for many students that means changing them.

Next, we offer two models to help you understand the skills and knowledge you will get from a quality engineering education, plus a third model to help you obtain that education.

We close the chapter by discussing the need for you to structure your life in ways that will minimize distractions and interferences. Only by doing so will you be able to devote adequate time to your studies and take advantage of the many resources available to you.

The material introduced in this chapter will provide a foundation for you to build on as you study the other chapters of this text.

1.1 You Can Do It!

From time to time I meet practicing engineers who tell me about the time when they were first-year engineering students and the dean told their Introduction to Engineering class:

> ***Look to your right, look to your left. Two of the three of you won't be here at graduation.***

It doesn't surprise me that engineering deans (and professors) say such upsetting things to students. They think that by scaring students about engineering studies, the students will be motivated to succeed.

What does strike me, however, is how angry these practicing engineers are at the dean for having given them such a negative message. And in some cases the event happened some 30 years before! These former students are still upset that the dean tried to frighten them at a time when they were unsure of themselves and easily intimidated.

When I meet with first-year engineering students, I convey a very different message. My message to them and to you is that:

> ***Each and every one of you can be successful in graduating with your bachelor of science degree in engineering.***

How can I make such a bold statement, without any specific information about your background or your ability? I'll tell you how.

POORLY PREPARED STUDENTS HAVE SUCCEEDED

For ten years I directed a program designed to enhance the academic success of engineering students. During that period I worked closely with more than 1,000 students. We had students with very poor preparation and limited ability—students who had to take college algebra three times before making a passing grade; students who failed trigonometry and had to repeat it, and then took Calculus I and made a *D* and had to repeat it. Some of those students took more than nine years of full-time study to complete their engineering degree.

I ran into one of those students many years later. He was a successful professional engineer and a respected member of his community. When I saw him, he was on the way to drop his daughter off at a relative's home so he could fly to Washington, D.C. for an important meeting.

HIGHLY QUALIFIED STUDENTS HAVE FAILED

I also worked with students who had all the preparation in the world—students who had gone to the best high schools and had excelled in their advanced mathematics and science courses. Yet they did not succeed in engineering study. Some flunked out. Some just dropped out.

The common denominator for such students was that they were overconfident. They had been able to excel in high school without a great deal of effort or need to adopt effective learning strategies. And they made the mistake of assuming that engineering study would be like high school. They naively believed that their ability would carry them through as it had before. They failed to account for the fact that the faster pace and higher expectations for learning would require substantially more effort and improved learning skills. And they didn't recognize that they had moved from the "minor leagues" to the "major leagues" where the ability of those they were competing against had increased substantially.

A few of those students have come back to see me. They express their deep regret for not sticking it out. It saddens me to hear they're working in unrewarding jobs for minimum salaries and would like to come back to school, but now the circumstances of their lives prevent them from having a second chance.

I hope you are not such a student. One early indication as to whether you are or not is how receptive you are to the material presented in this book. Thinking there is nothing of value here for you is a sign that you are overconfident. If you are, I hope you will consider this section as a "wake-up call." You can ignore this warning with the intent of shifting gears later. The problem with that approach is that your early courses, particularly in mathematics and science, provide the foundation on which your entire engineering education will be built. If you start out with a weak foundation, you will find it very difficult, if not impossible, to build a sound structure on top of it.

REFLECTION

In the previous two sections, we talked about "Poorly prepared students have succeeded" and "Highly qualified students have failed." Do you see something of yourself in either category? Do you lack confidence? If so, are you beginning to believe you can do it? Or are you overconfident? If so, are you beginning to become receptive to learning new strategies and approaches for your engineering studies?

WHAT MAKES THE DIFFERENCE?

One student with seemingly limited ability and poor preparation succeeds. Another student with outstanding ability and excellent preparation fails. How can that happen? What are the keys to success in engineering study? What are those things you can do that will virtually ensure that you succeed—those things that if not done will at best result in your working below potential and could even lead you to failure?

Success in engineering study is not unlike success in anything you have attempted or will attempt. Achieving success is a process, and each step in the process can be learned. I would encourage you to make a commitment to become an "expert" on success. It's something that you can do. And the payoff will be enormous.

Lots of resources are available to help you. You can learn from others, from reading books, from listening to audiotapes, from the wealth of information available on the Internet, and from attending short courses and workshops. Many of the best resources are listed at the end of each chapter. Make learning about success one of your hobbies. If you work at it, your capacity to be successful will expand and grow throughout your life. You might even surprise yourself at what you can achieve. And who knows? Maybe someday you'll write a book on *success* for others.

1.2 WHAT IS "SUCCESS"?

I assume that you want to be successful. Otherwise you wouldn't be reading this book. But just wanting to be successful is not enough. Everyone wants to be successful. Often when speaking to an Introduction to Engineering class, I'll ask the question, "How many of you want to be successful?" All of the students raise their hands. But what do the students mean when they say they want to be successful? Are they all thinking about the same thing? Probably not.

When I ask the same students, "What is *success*?" I get a variety of answers:

- *Success* is being happy.
- *Success* is making money.
- *Success* is having control over your life.

But almost always one or more of the students will give the right answer:

- *Success* is the achievement of goals.

Webster's Dictionary says essentially the same thing:

> ***Success* is the achievement of something desired, planned, or attempted.**

The point is that unless you have something "desired, planned, or attempted," there can be no success.

Unfortunately, many students lack a clear goal and commitment to that goal necessary for success. According to Vincent Tinto [1], author of an excellent book on student success, the top two reasons why students do not succeed in college are:

(1) Lack of Intention - Students do not have a clear educational and/or career goal.

(2) Lack of Commitment - Students do not have the motivation and drive to work toward attaining their educational/career goals.

Identifying a clear goal and developing a strong commitment to that goal are the essential first two steps in the process of achieving success.

REFLECTION

Reflect on the relationship between *success* and *happiness*. What does each of these words mean? Does success bring happiness? Can people be happy if they are not successful? Think about Dale Carnegie's quote at the beginning of this chapter: *Success is getting what you want; happiness is wanting what you get*. Do you usually get what you want? Do you usually want what you get?

GOAL SETTING

If success requires a goal, let's discuss goal setting. Obvious though it may sound, the basic idea behind goal setting is:

> ***How can you ever expect to get somewhere if you don't know where you want to go?***

That is, setting goals—having a specific idea of what you want to accomplish in both the short and long term—is a key requirement to becoming an effective student and a successful professional. Only when you set goals will you have something to strive for and something against which to measure yourself.

GOALS GIVE YOU SOMETHING TO MEASURE YOURSELF AGAINST. Consider, for example, two engineering students in a calculus class who score a *B* on their first exam. One student is extremely unhappy and resolves to study much harder for the next exam. She has set a goal of earning an *A* in the course and by falling short on the first exam, she knows that she must work more. The other student, however, is content with the *B* grade and decides that he can increase his work hours since even less study is necessary than he thought.

These different behaviors are the result of different expectations—of the two students having different goals for themselves. As this case illustrates, success or failure can only be measured according to self-imposed goals.

GOALS GIVE YOUR LIFE DIRECTION. I'm sure you were asked many times during your childhood, "What do you want to be when you grow up?" If you didn't know, you probably felt a bit frustrated and even irritated at people who asked you that question. But I hope you realize by now that they were trying to help you. They were trying to alert you to the importance of setting directions for your life. Doing so may not be easy, but the payoff is definitely worth the effort, as the stories of many successful people indicate. Following is but one such story.

Astronaut Franklin Chang-Diaz Story

Dr. Franklin Chang-Diaz is one of the most accomplished astronauts at NASA. A veteran of six space missions, he has logged over 1,269 hours in space. But when you hear the story of his life, you wouldn't think he'd end up in such a prestigious position.

Chang-Diaz was born and grew up in Costa Rica. As a child, he was enamored of the U.S. space program. He and his friends used to build spacecrafts out of cardboard boxes, equipping them with broken radios, furniture, and other discarded material. They would then go through a countdown and lift-off and pretend to travel to distant planets. Because of his interest, Chang-Diaz set a personal goal of becoming a U.S. astronaut. Imagine a young Costa Rican citizen, who didn't speak a word of English, aspiring to be a U.S. astronaut!

When he finished high school, he worked for a year and saved enough money to buy a one-way airplane ticket to Hartford, Connecticut, where he had some distant relatives. In Hartford he repeated his senior year of high school, learned English, and was admitted to the University of Connecticut, where he majored in engineering. After graduating with honors, he began graduate study at MIT, eventually receiving his Ph.D. in plasma physics. He then applied for the astronaut program, was accepted, and became the U.S.'s first Hispanic astronaut.

To learn more about Dr. Chang-Diaz, visit his NASA web site at: http://www.jsc.nasa.gov/Bios/htmlbios/chang.html.

The point that the story of Dr. Chang-Diaz drives home so convincingly is the need to have goals. Hearing his story makes me wonder what I might have accomplished had I set such lofty goals.

Astronaut Franklin Chang-Diaz

<u>Write Down Your Goals</u>. Right now your primary goal should be to graduate with your degree in engineering. But what else would you like to accomplish? Become president of your own company? Become a multimillionaire? Become a college professor? And what about your more immediate goals? Maybe you want to make a 3.0 GPA next semester, improve your writing skills, or become president of one of the engineering student organizations.

A good exercise would be for you to write down your short-term, intermediate-term, and long-term goals. Consider what you want to accomplish in the next week, in the next month, in the next year, in the next five years. Review and update these lists regularly.

> ***Start by making Graduation in Engineering one of your primary life goals.***

STRENGTHENING YOUR COMMITMENT

Why did you choose engineering as your major? Perhaps because you were good in math and science, one of your high school teachers or counselors recommended that you study engineering. Or maybe you are doing it to please your parents, or because you don't know what else to do. It is likely that you don't know a great deal about engineering. Few students do.

Regardless of your reasons for being here, it is critically important that you develop a strong motivation to succeed. Engineering is a demanding field of study. Even a student with excellent preparation and strong ability will not succeed without a high level of commitment.

There are at least three practical strategies you can use to strengthen your commitment to success in engineering study:

(1) Clarifying your goals

(2) Learning as much as you can about engineering

(3) Developing a "road map"

CLARIFYING YOUR GOALS. What does it mean to clarify your goals? Very simply, it means answering such questions as, "Why do I want to achieve the goal?" "What will it mean to the quality of my life if I am successful in accomplishing the goal?" Clarifying your goals helps you understand their value to you personally. And by better understanding their value, you will become more committed to achieving them.

As noted earlier, many students know very little about what engineering is and what engineers do. In particular, they tend not to know about the tremendous rewards and opportunities that an engineering degree offers. Learning about these rewards and opportunities, as we will do in Chapter 2, will figure significantly in clarifying your personal goals.

LEARNING AS MUCH AS YOU CAN ABOUT ENGINEERING. As you have grown up, you have been exposed to teachers, doctors, dentists, ministers and rabbis, and numerous other professionals. You have a feel for what accountants do if you have had to manage your personal finances. You have seen lawyers at work on TV shows such as *Law and Order* and *Boston Legal.* Through your coursework, you have developed some feel for what mathematicians, chemists, and physicists do. It is doubtful, however, that you have had much exposure to engineering. The exposure you have had has probably been indirect, through contact with the products that engineers design.

Learning about engineering is a lifelong process, but it should begin now. Take advantage of every opportunity that presents itself. You can start by studying Chapter 2 of this text thoroughly. Explore some of the many Internet web sites referred to there, particularly those whose purpose is to help students learn about engineering. Attend seminars on career opportunities, go on field trips to industry, talk with company representatives at career day programs. Browse the resource library in your career center. Become active in the student chapter of the professional engineering society for your major. Talk to your professors. Read biographies of successful engineers [2, 3, 4, 5]. If you land a summer job in industry, be curious and inquisitive. Look around. Talk to the engineers and find out what they do.

Over time, these efforts will pay off and your understanding of engineering will increase. Increased knowledge will bring increased motivation. We tend to like things we know a lot about.

PREPARE A "ROAD MAP" Remember when you were in the 5th grade, and you heard the term "algebra" and thought, "I'll never be able to learn that!" And later you were overwhelmed with the thought of mastering trigonometry or calculus. You thought you wouldn't be able to handle such advanced subjects, but you were wrong. Each time you reached the next higher level, you were able to handle it, even excel at it. How did you do it? By taking lots of little steps, each one building on previous steps.

Often students ask me, "What does it take to succeed in engineering study?" My answer is, "You must be able to pass Calculus I at the university level." My reason for this is very simple. If you can pass Calculus I, you can pass Calculus II. And if you can pass Calculus II, you can pass Calculus III. If you can pass Calculus III, you can then pass Calculus IV. And if you can pass these calculus requirements, you can pass the junior engineering courses. If you can pass the junior engineering courses, you can pass the senior engineering courses.

So you see, succeeding in your engineering program is a process of taking one little step after another. Progressing through the engineering curriculum is just an extension of what you have already demonstrated you can do.

I suggest you develop a "road map" that will lead you to graduation in engineering. Lay out a plan of what you will need to take each semester or quarter to complete your engineering program. Having a step-by-step road map to follow will progressively increase your confidence and strengthen your commitment to achieve your ultimate goal: that B.S. degree in engineering.

DON'T LET ADVERSITY STOP YOU. Highly successful football coach Lou Holtz, a sought after motivational speaker, relates a primary difference between people who succeed and people who fail. According to Holtz:

> ***People who succeed are people who when they get "knocked down" by some adversity, they get up; whereas, people who fail are people who when they get knocked down, they stay down.***

I would encourage you to read Coach Holtz's recent autobiography *Wins, Losses, and Lessons* [6].

The most likely reason you will fail to graduate in engineering is that you will encounter some adversity and give up. You will have difficulty with a course or with a professor. Or you will have a personal problem, a relationship problem, or a health problem. Whatever adversity you are bound to experience, you will be tempted to use it as an excuse or justification for quitting. Don't!

By strengthening your commitment following the steps outlined in the previous three sections, you will develop "determination." The dictionary defines *determination* as "a firmness of purpose . . . having one's mind made up." Determination means having an unwavering commitment to your goal—the goal of graduating in engineering. You must be determined to persist, particularly in the face of adversity.

A Personal Story

I dropped out of college early in my sophomore year. I had learned as I attempted to register for my second year that I had lost my full tuition scholarship because of poor grades. Faced with taking out a massive loan and having broken my leg playing intramural football, I dropped out. I had always wanted to be a jet pilot anyway, so as soon as my cast was off, I went directly to the local Air Force Recruiting Office. To my chagrin I was told a college degree was required for flight training. Soon I was back in school with newfound determination. That experience was a significant lesson to me that doors would be shut without a college education.

Adopt the view that you are going to achieve your goal and that nothing is going to stop you. And how do you keep adversity from stopping you? How can you keep failures from discouraging you? I find the age-old saying:

We learn more from our failures than we do from our successes.

to be very helpful as a philosophical basis for overcoming adversity. It's true! Think about it.

Another Personal Story

When I was in the 7th grade, I took a gymnastics class. I was the best in the class on the side horse. So when we had a competition at the end of the term, everyone just knew that I would win that event. But when I began performing, I was so nervous I felt like needles were pricking my skin all over. I came in last. I was terribly embarrassed and ashamed. It took me a long time to get over that failure. But that experience showed me that if I take myself too seriously and want to win too much, I can actually perform much worse than I am capable of. That experience has helped me deal effectively with high-pressure situations ever since.

Learning to overcome adversity as a student will also benefit you during your professional career. Joseph J. Jacobs, founder and former CEO of Jacobs Engineering and one of the nation's most successful engineers and businesspersons, gives his "Nine Commandments for the Entrepreneur." The first four are:

(1) You must be willing to risk failure.

(2) You must passionately hate failure.

(3) Persistence is a necessity, just as is the willingness to acknowledge defeat and to move on.

(4) A measure of your potential to succeed is how you handle adversity.

(I encourage you to read Mr. Jacob's highly motivational autobiography, *The Anatomy of an Entrepreneur* [5].)

REFLECTION

Do you believe the statement, "You learn more from your failures than you do from your successes"? Have you ever experienced a significant failure? What was it? What did you learn from that experience?

If you are determined to graduate in engineering, if you persist even in the face of adversity, if you take the view that you will not allow anything to stop you, the chances are very good that you will succeed.

Believe in yourself. You can do it!

1.3 KEYS TO SUCCESS IN ENGINEERING STUDY

Setting a goal and making it important to you are only the first steps. The real challenge remains—achieving the goal. Once your goal is identified and you have done everything you can to develop a strong commitment to that goal, achieving it requires that you adjust both your attitudes and your behaviors to those appropriate to the goal. This means that you make your day-to-day decisions and choices based on whether a particular action supports your goal (i.e., moves you closer to your goal) or conflicts with your goal (i.e., moves you farther away from it).

In my experience there are three keys to success in engineering study:

Effort - ***Work hard***

Approach - ***Work smart***

Attitude - ***Think positively***

Let's examine each of these.

EFFORT - "WORK HARD"

Do you believe that **people succeed because of their ability**, that some people "have it" while others don't? Or do you believe that **people**

succeed because of their effort? An excellent book that contrasts these two ways of looking at the world is *Mindset: The New Psychology of Success* by Carol S. Dweck [7].

The first belief—that some people have it and some don't—is a self-defeating belief. Thinking that you don't have as much ability as others can provide you with a rationale to accept personal failures. You may as well give up. After all, if success is related to some natural quality that you have no control over, then it doesn't matter what you do.

Believing you're the smartest kid on the block has pitfalls as well. If you do, you're likely to feel that you have to prove yourself over and over, trying to look smart and talented at all costs. And research has shown that people with this mindset are more likely to stick with approaches that clearly don't work, and ignore suggestions from others.

The second belief—that people succeed because of their effort—is empowering because the amount of effort you put in is in your direct control. You can choose to put in more effort and in doing so significantly affect your success.

ABILITY VS. EFFORT. The relative importance of ability and effort was perhaps best put by the famous American inventor Thomas Edison:

> ***Genius is one percent inspiration and 99 percent perspiration.***

Does the following dialogue sound familiar to you? Over the years, I've had a variation of it with many of my students.

Landis: How's everything going?

Student: *Fine!*

Landis: What's your hardest course this term?

Student: *Physics: Electricity and Magnetism.*

Landis: How are you doing in that course?

Student: *Fine!*

Landis: What score did you make on the last exam?

Student: *Forty-three.*

Landis: What grade is that?

Student: *I don't know.*

Landis: Is it an "A"?

Student: *No*

Landis: A "B"?

Student: *No.*

Landis: A "C"?

Student: *Probably not.*

Landis: A "D"?

Student: *Maybe.*

Landis: 'Sounds like an "F" to me. How many hours are you putting into your physics course?

Student: *About 15 hours a week.*

Landis: How many hours have you studied today?

Student: *I haven't done any studying today.*

Landis: How many hours did you study yesterday?

Student: *None yesterday.*

Landis: How about over the weekend?

Student: *I meant to, but just never got to it.*

Landis: So you're planning to study physics for five hours a day for the next three days to get your 15 hours in this week?

EFFORT IS BOTH TIME AND ENERGY. In my experience, poor academic performance can usually be traced to insufficient effort. Just what do I mean by "effort"? It is "using energy, particularly mental power, to get something done."

The effort you devote to your studies has two components—time and energy. An analogy can be made using the well-known physics formula:

Distance = Rate x Time

Completing a specific task (i.e., traveling a distance) requires that you devote energy or mental power (rate) and spend time on the task (time). In later sections, we will consider how much time is sufficient, what is the best use of that time, and when to put in that time if you want to be both effective and efficient.

The important point here is that your success in the study of engineering is to a great extent **in your control**. How well you perform will depend, in large measure, on how much effort you put in. Accomplishing an academic task, like completing a homework assignment, will require you to devote adequate time and to focus your energy and mental power. These are things that you can choose to do or choose not to do.

APPROACH - "WORK SMART"

"Approach" refers to ***how*** you go about your engineering studies. It means that you work not only hard but "smart." To a great extent, your approach to your engineering studies depends on the ideas we have already discussed. It assumes that:

- You know why you want to be an engineer and appreciate the value of a technical education.
- You have clarified your goals and developed a "road map" to lead you to them.
- You are strongly committed to achieving your goals, even in the face of adversity.
- You have gotten your life situation together, so that you are not overburdened with problems and distractions—and if so, you are prepared to make the necessary choices and personal sacrifices.

Above all, however, your approach to your engineering studies—working "smart"—means that you learn to become a ***master*** engineering student.

BECOMING A MASTER STUDENT. To understand what I mean by becoming a *master* student, consider the following analogy. If you were to take up chess, what would you do? Learn the basic objectives, rules, and moves and then begin to play? Probably. But you'd soon discover that mastering a game of skill like chess requires much more. So you might read a book, take a lesson, or watch experts play. You would realize that to become a chess *master*, you need to spend time both playing the game and learning about it.

Your approach to the study of engineering can be likened to a game. To become a *master* student, you must not only play the game—i.e., be a student; you must also devote time and energy to learning how to play it.

The first step in playing the "game" of becoming a master engineering student is to get a clear picture of what is required to earn your B.S. degree. Earlier, when discussing what it means to prepare a "road map" for yourself, I gave a brief synopsis of what you need to do to graduate in engineering. Let me give you a related description here: You become an engineer when you pass a set of courses required for an engineering degree. What is required to pass each course in the set? Primarily passing a series of tests or exams. And to pass the series of tests, you must pass each test one at a time. So by breaking it down this way, you can see that to become an engineer, you must become a master at preparing for, taking, and passing tests.

Of course, this is easier said than done, because many other factors are involved. But by approaching your engineering studies in this light, the "game" of becoming a master student and, ultimately, earning your engineering degree becomes less daunting.

As you read the subsequent chapters in this book, you will discover different ideas and perspectives on how best to approach your studies. Learning to be a master engineering student will be a tremendously rewarding and beneficial experience. It will enhance your immediate success as a student, while developing important skills you will later need as a practicing professional engineer. Indeed, many of the approaches you learn in this book will work for you in whatever you do.

ATTITUDE - "THINK POSITIVELY"

Are you a positive person? Or are you a negative person? Are you aware of the role attitude plays in your success? What do you think of the following statement?

Positive attitudes produce positive results.
Negative attitudes produce negative results.

Among those negative attitudes that could produce negative results in engineering study are:

- Weak commitment to the goal of graduating in engineering
- Low self-confidence
- Unrealistic view of what's expected to succeed in engineering studies (overconfidence, naiveté)

- Lack of self-worth (i.e., tendency to sabotage your success)
- External "locus-of-control" (i.e., adopting a "victim" role)
- Unwillingness to seek help (e.g., thinking that seeking help is a sign of weakness)
- Resistance to change (e.g., your behaviors and attitudes)
- Tendency to procrastinate (e.g., having a negative view about the idea of managing your time)
- Avoidance of areas of weakness or perceived unpleasantness (e.g., writing, oral presentations, difficult courses)
- Reluctance to study with other students (e.g., avoidance of group study)
- Negative view toward authority figures (e.g., parents, professors)

REFLECTION

Think about each of the negative attitudes in the list above. Do any of the items describe you? If so, in what ways could you see that particular attitude interfering with your success in engineering study? Do you know why you hold this attitude? Are you willing to try and change the attitude? What would be a more positive attitude that you could adopt?

One of the primary purposes of this book is to help you become conscious of and change any negative attitudes you may hold that will impede your success in engineering study. You will learn the process for this change when you study Chapter 6: *Personal Growth and Development*.

1.4 MODELS FOR VIEWING YOUR EDUCATION

One of the most positive and unique aspects of your college experience is that you are working for yourself to prepare yourself for your future. Consider the saying:

No deposit, no return

Your education represents a significant deposit, or investment, you are making in yourself. Your return will be in direct relation to what you put in. You must realize that whenever you take the easiest instructor, avoid a tough course, or cut a class, you are hurting yourself. Whenever you make a conscious choice to avoid learning, growing, or developing, you are not getting away with something—**you are working against yourself!**

If you want to get the most out of your education, you need a model from which to view it. Earlier in this chapter, I gave simplified explanations of the engineering curriculum in order to demystify it for you. First, I described it as a required set of courses that you must take. Later, I broke down each course as a series of exams you must pass.

It is time now to broaden your view of your engineering studies, because a quality education involves much more.

The purpose of the next three sections is to give you three models from which to view your education. These models will assist you in answering such important questions as:

- What is the purpose of my education?
- What should I know when I graduate?
- How do I know if I am getting an excellent education?
- How can I enhance the quality of my education?
- Will I have the knowledge and skills to get a job?

These models are also useful for personal assessment or self-evaluation. My suggestion is that you measure yourself against each item presented in these models. In other words, ask yourself on a scale of 0 to ten (ten being highest): *How would I rate myself on this item?* In areas you feel you are strong, just keep doing what you have been doing. In areas you need to improve, map out a plan to strengthen these areas. Personal assessment and personal development plans will be discussed in more detail in Chapter 6.

ATTRIBUTES MODEL

In today's tight fiscal climate, universities are being held more accountable for their productivity. Institutions are being asked to establish educational objectives and student outcomes and to show that these objectives and outcomes are being met. This process is called *institutional assessment*. It is not unlike what happens to you in your classes. Your professor sets certain course objectives and has certain expectations of how well you will do in achieving these objectives. At the end of the term, the degree to which you meet these expectations is measured and transmitted to you in the form of a final grade.

One way engineering programs are held accountable is through the accreditation process administered by the Accreditation Board for Engineering and Technology (ABET). Understanding the accreditation process (which is discussed in more detail in Chapter 8) will help you better understand the engineering education you are beginning.

ABET, through its *Engineering Criteria 2000* [8], mandates that engineering programs must demonstrate that their graduates have the following 11 attributes:

ABET Attributes of Engineering Graduates

a. An ability to apply knowledge of mathematics, science, and engineering
b. An ability to design and conduct experiments, as well as to analyze and interpret data
c. An ability to design a system, component, or process to meet desired needs
d. An ability to function on multi-disciplinary teams
e. An ability to identify, formulate, and solve engineering problems
f. An understanding of professional and ethical responsibility
g. An ability to communicate effectively
h. A broad education necessary to understand the impact of engineering solutions in a global and societal context
i. A recognition of the need for, and an ability to engage in, life-long learning
j. A knowledge of contemporary issues
k. An ability to use the techniques, skills, and modern engineering tools necessary for engineering practice

This list of attributes provides you a clear picture of what you should get from your engineering education. That is, when you complete your engineering degree, you will have the knowledge, skills, and attitudes you will need for a successful and rewarding career.

REFLECTION

Reflect on each of the 11 attributes of engineering graduates required by ABET. To the extent possible, think about what is meant by each item on the list. Reflect on the perspective that this is virtually a "blueprint" for what you should gain from your engineering education. Consider which areas are most appealing to you. Which areas are you likely to excel in? Design? Communication skills? Teamwork? Problem solving? Use of engineering tools? Experimentation? Ethical responsibility? Other?

EMPLOYMENT MODEL

A second model that may be useful to you in viewing your education is the *Employment Model*. Certainly, one reason why many students choose to major in engineering is the availability of jobs. In light of this, you need to consider what characteristics are important to employers, and work to develop yourself in these areas. In study after study, employers consistently rank the following as the top six factors when considering individuals for employment:

(1) Personal qualifications including maturity, initiative, enthusiasm, poise, appearance, integrity, flexibility, and the ability to work with people

(2) Scholastic qualifications as shown by grades in all subjects or in a major field

(3) Specialized courses relating to a particular field of work

(4) Ability to communicate effectively, both orally and in writing

(5) Kind and amount of employment while in college

(6) Experience in campus activities, especially participation and leadership in extracurricular life

As you approach graduation, you will undoubtedly participate in a number of interviews with prospective employers. How you fare in those interviews will depend largely on how well you prepare yourself between now and then in the six areas listed above. To be strong in each area, you must make a conscious commitment to make it happen.

Subsequent chapters in this book offer guidance and suggestions to help you acquire these attributes.

- **Chapters 3, 4, and 5** will address academic success strategies that will ensure you have strong scholastic qualifications.
- **Chapter 6** will instruct you in ways to develop your personal qualifications.
- **Chapter 7** will explain the value of active involvement in student organizations and engineering-related work experience.

STUDENT INVOLVEMENT MODEL

Let's assume that you want to get a *quality* education—i.e., to acquire the knowledge, skills, and attitudes that will result in your being highly sought after by engineering employers. How can you guarantee that you get that *quality* education? In fact, what do we mean by "quality" or "excellence" in education? We can find the answer in a paper entitled "Involvement: The Cornerstone of Excellence" by Alexander W. Astin, Director of UCLA's Center for the Study of Higher Education [9].

GETTING AN EXCELLENT EDUCATION. According to Astin, an "excellent" education is one that maximizes students' intellectual and personal development. He says the key to students' intellectual and personal development is a high level of "student involvement." Astin defines student involvement as:

> ***the amount of physical and psychological energy that the student devotes to the academic experience.***

And he gives five measures of student involvement:

(1) Time and energy devoted to studying

(2) Time spent on campus

(3) Participation in student organizations

(4) Interaction with faculty members

(5) Interaction with other students

Put simply by Astin:

> *A **highly involved student** is one who, for example, devotes considerable energy to studying, spends a lot of time on campus, participates actively in student organizations, and interacts frequently with faculty members and other students.*

Conversely, according to Astin:

> *An **uninvolved student** may neglect studies, spend little time on campus, abstain from extracurricular activities, and have little contact with faculty members or other students.*

REFLECTION

Evaluate yourself against Astin's five measures of student involvement. Would you describe yourself as a "highly involved student"? Or would you describe yourself as an "uninvolved student"? Do you agree that Astin's five measures relate to the quality of the education you are receiving? Are you willing to make changes to take advantage of Astin's model?

INVOLVEMENT IS UP TO YOU. The Astin "student involvement" model suggests that the quality of the education you get will depend primarily on the approach you take to your studies. Although your institution can do things to encourage you to study more, to spend more time on campus, to become involved in student organizations, to interact with your professors, and to interact with fellow students, increasing your level of involvement is mostly up to you.

You can choose to devote more time and energy to your studies, to spend more time on campus, and to become active in student organizations. You can choose to interact more with your professors and to become more involved with other students. In doing so, you will greatly enhance the quality of your education.

1.5 STRUCTURE YOUR LIFE SITUATION

I hope that the ideas presented thus far in this chapter have convinced you of the importance of making success in engineering study one of your primary life goals and have strengthened your commitment to that goal. One of the key objectives of Chapter 2 will be to further strengthen that commitment by increasing your understanding of engineering as a profession and giving you a clear picture of the rewards and opportunities that an engineering career offers you. With a clear goal and a strong commitment to it, you are well on your way to achieving that goal. All that remains is to do it.

The first step in "doing it" is to create a life situation that supports your goal. Full-time engineering study is a major commitment, so you must be prepared to devote most of your time and energy to it. This means eliminating or minimizing any external distractions or obligations that will interfere with your studies—and work against your goal.

I often encounter students who are taking a full load of math/science/ engineering courses while commuting over an hour each way to school, working 20 or more hours per week, responding to demands placed on them by their family, and trying to maintain an active social life. Students in such situations are very likely programmed for failure.

Whether demands outside of school come from family, friends, work, commuting, or any other source, you need to make whatever changes are necessary so that you, too, don't program yourself for failure.

Living Arrangements

If at all possible, live on or near the university campus. The more immersed you can get in the university environment, the better your chances of success will be. Commuting takes time, energy, and money; and living at home can present problems. Parents may expect you to help with the household duties. Little brothers and sisters may be noisy and distracting. Neighborhood friends may not understand your need to put your studies ahead of them. Wherever you live, however, remember that now is a time in your life when it's appropriate to be a bit selfish. Place a high value on your time, and learn to say no when necessary.

Regardless of your living arrangement—at home with parents; in an apartment alone or with a roommate; or in an on-campus residence hall—I would encourage you to come to campus early in the day and do your work there, rather than just come to take classes and leave as soon as possible. Your university or college campus is an "academic place." Its primary purpose is to facilitate the teaching/learning process. And it's set up to do just that. Whereas, at home, apartment, or residence hall there are many distractions (TV, stereo, telephone, refrigerator, friends, parents, siblings), on campus there are lots of resources (professors, other students, places to study, library resources, tutors). I would encourage you to approach your engineering study much like you would a full-time job in that you go to your "place of work" and do the greater share of your work there, perhaps bringing some work home, but certainly not all.

Sometimes you'll hear the viewpoint that students at so-called "residential campuses" get more out of their education than students at so-called "commuter campuses." Putting the approach outlined in the previous paragraph into operation, in effect, brings all the benefits of being a residential student to a commuting student.

PART-TIME WORK

As noted above, full-time engineering study is a full-time commitment. Working up to ten hours a week at a part-time job is probably okay, but more is almost certain to take its toll on your academic performance. While it may be essential for you to work, it may also be that you are working to afford a nice car, expensive clothes, or other non-essentials. Look at it this way. You may get a job for $8-10 an hour now, but in doing so you jeopardize your education or at best extend the time to graduation. The average starting salary for engineering graduates is around $24 an hour. If your career is successful, someday you might make more per hour than students make per day. So try to delay as many material wants as possible. By doing so, you will have much more in the long term.

If you must work while going to school, particularly if your work exceeds ten hours per week, how can you achieve a reasonable balance between the two? One way is to follow the guidelines below.

Hours worked	Max. course load
10 hrs/wk	full load
20 hrs/wk	12 units
40 hrs/wk	8 units

Another way to manage your study and work loads is to follow the "60-Hour Rule" espoused by Dr. Tom Mulinazzi, Associate Dean of Engineering at the University of Kansas [10]. This is not a rigid rule, but rather a guideline. It doesn't apply to a single week, but is a pretty good rule-of-thumb over the long haul.

The 60-Hour Rule

The "60-Hour Rule" is an excellent rule to follow. I have shared it with freshmen in an Introduction to Engineering course each fall.

The Rule is stated as follows: It is assumed that a student can "work" for 60 hours a week over the period of a term. This work includes academic work, work at a paying job, and commuting time. The Rule also assumes that a student must study two hours for every hour in the classroom.

Let's say that a student is working 20 hours on campus. Take 60 less 20, and the result is 40. Divide 40 by three (one hour in class and two hours of studying per week for every credit hour) and the result is 13. This means that most students can take 13 units of coursework and derive satisfactory results while working 20 hours. Ninety-five percent of the engineering students who are dismissed from the University of Kansas School of Engineering violate the 60-Hour Rule.

From time to time, I encounter a student who is taking four courses each term but passing only two of them. When I suggest that the student reduce his or her course load, the typical response is, "I can't do that. It'll take me forever to graduate!" Obviously, though, such students are moving through the curriculum at the rate of only two courses per term. The point is, be realistic about your situation. Don't create an unmanageable workload and then deceive yourself into thinking that it is working.

INFLUENCE OF FAMILY AND FRIENDS

Because family and friends may not understand the demands of engineering study, they may unintentionally distract you. If your family poses problems, have a frank talk with them. Let all family members know that you want to make school your #1 priority. Ask for their help, and negotiate clear agreements about their demands on you.

If you are a recent high school graduate, dealing with friends from high school—especially those who are not pursuing a college education—may be difficult. These friends may put pressure on you to spend as much time with them as you did in high school, while you may find that you not only don't have time for them, but that you also have less and less in common with them.

If you find yourself in this situation, you alone will have to decide how to handle it, as there are no easy answers. But it is important that you be realistic and that you understand the consequences of your choice to study engineering. By making this choice, you are moving yourself in a different direction that may increasingly distance you from your old friends, while bringing you into contact with new people and peers—and opportunities for new friendships.

However you decide to deal with your old friends, by all means do not let them keep you from the opportunities to develop new friendships at school. I can't encourage you enough to cultivate relationships with your fellow engineering students, for befriending them will be tremendously rewarding. Not only will you likely be initiating important lifelong relationships; you also will derive the immediate benefits of being able to integrate your academic and social lives, while building a support system in which "friends help friends" to achieve the same academic goals.

REFLECTON

Think about your life situation in terms of the factors presented in the previous sections. Are there things that come to mind about your living arrangements, your work-load, or the influence of family and friends that need to be changed if you are going to be successful in engineering study? List those things and develop some first steps you can take to bring about this needed change.

SUMMARY

This chapter introduced you to the keys to success in engineering study. We first focused on the importance of making graduation in engineering your primary goal at this time in your life.

Next, we presented three strategies for strengthening your commitment to that goal: (1) clarifying why you want to be an engineer; (2) learning as much as you can about engineering; and (3) developing a step-by-step guide, or "road map," that you can follow.

We noted that achieving a goal requires you to adopt appropriate attitudes and behaviors. We also discussed the importance of effort, in terms of both time-on-task and energy (i.e., mental power). Last, we explored the importance of the approach you take to your engineering studies. We saw that success not only means that you study "hard" but also that you study "smart."

Three models were then presented to help you understand what a "quality" education entails.

(1) The first model listed the attributes all engineering graduates must have as mandated by the Accreditation Board for Engineering and Technology (ABET).

(2) The second model focused on the qualifications that employers seek when considering candidates for entry-level engineering positions.

(3) The third model stressed the importance of "student involvement" to ensure that you get a "quality" education.

Each of these models identifies the knowledge, skills, personal qualities, and behaviors that you need to develop during your college years. Each model also provides specific areas against which you can assess yourself. Doing periodic personal assessments will point out your strengths and areas for improvement.

We closed the chapter by talking about the need to structure your life situation so that it supports your goal of graduating with an engineering degree. The gist of this discussion centered on your ability to balance the demands of your school work with outside demands—from jobs, family, friends, and all other sources—so that you reserve adequate time to devote to your studies.

REFERENCES

1. Tinto, Vincent, *Leaving College: Rethinking the Causes and Cures of Student Attrition, Second Edition*, The University of Chicago Press, Chicago, 1993.

2. Iacocca, Lee, *Iacocca: An Autobiography*, Bantam Books, New York, 1986.

3. Hansen, James R., *First Man: The Life of Neil A. Armstrong*, Simon & Schuster, 2005.

4. Hickam Jr., Homer H., *Rocket Boys*, Delta Publishing, 2000.

5. Jacobs, Joseph J., *The Anatomy of an Entrepreneur*, ICS Press, Institute for Contemporary Studies, San Francisco, 1991.

6. Holtz, Lou, *Wins, Losses, and Lessons: An Autobiography*, William Morrow, 2006.

7. Dweck, Carol S., *Mindset: The New Psychology of Success*, Random House, 2006.

8. Accreditation Board for Engineering and Technology (ABET), 111 Market Place, Suite 1050, Baltimore, MD 21202 (*Engineering Criteria 2000* available on ABET web page: http://www.abet.org)

9. Astin, Alexander W., "Involvement: The Cornerstone of Excellence," *Change*, July/August 1985.

10. Mulinazzi, T., "The 60-Hour Rule." *Success 101, Issue 1*, Spring, 1996. (Available from: R. B. Landis, California State University, Los Angeles, Los Angeles, CA 90032).

PROBLEMS

1. Have any of your teachers or professors ever done anything to make you feel as though you couldn't make it? What did they do? Why do you think they did that?

2. Discuss the relationship between *success* and *happiness*. What does each of these words mean? Does success bring happiness? Can people be happy if they are not successful?

3. Do you have a personal goal of graduating with your bachelor of science degree in engineering? How important is that goal to you? How can you make it more important?

4. Develop a list of 20 goals you would like to accomplish in your lifetime. Be bold!

5. Establish a goal for the grade you want to achieve in each of your courses this term. What GPA would this give you? How would it compare to your overall GPA?

6. List ten benefits that will come to you when you're successful in graduating in engineering. Rank them in order of importance to you.

7. List ten tasks that an engineer might perform (e.g., write a report, conduct a meeting, perform a calculation). Rank them in the order that you would most enjoy doing. Explain your ranking.

8. Read a biography of a famous engineer. Write a critique of the book. Include a discussion of what you learned from the book that will help you succeed in engineering study.
9. Do you believe the statement, "You learn more from your failures than you do from your successes"? Have you ever experienced a significant failure? What was it? What did you learn from that experience?
10. Have you ever achieved anything that others thought you couldn't through sheer determination? What was it?
11. How many hours do you think you should study for each hour of class time in your mathematics, science, and engineering courses? Is this the same for all courses? If not, list four factors that determine how much you need to study in a specific class.
12. Ask one of your professors why he or she chose teaching as a career rather than professional practice.
13. Would you rather tackle an easy problem or a difficult one? Which do you think benefits you more? Make an analogy with the task of developing your physical strength.
14. List five things you could do to study "smart" that you are not currently doing. Pick the two most important ones and try to implement them. Prepare a brief oral presentation for your Introduction to Engineering class that discusses your success or lack of success in implementing them.
15. List six things that your professors can do for you beyond classroom instruction.
16. If you spend 100 hours studying, how many of those hours would you be studying alone? How many would you be studying with at least one other student? If you study primarily alone, why? List three benefits of working collaboratively with other students.
17. Check off any of the statements below that describe your attitude.

ATTITUDE

My commitment to success in engineering study is weak.	
I lack confidence in my ability to succeed in engineering study.	
I have a tendency to sabotage my success.	

I tend to blame others for my failures.	
I don't see any need to change myself or to grow or develop.	
I am generally unwilling to seek help from others.	
I tend to procrastinate, putting off the things I need to do.	
I tend to avoid doing things that I don't enjoy.	
I avoid contact with my professors outside of class.	
I prefer to study alone rather than with other students.	

For any of the items you checked, answer the following questions:

a. Is this attitude working for me (positive attitude) or working against me (negative attitude)?

b. If the attitude is working against me, can I change it? How?

18. Rank ABET's list of 11 attributes of engineering graduates presented in Section 1.4 in order of importance. Meet with your engineering advisor or an engineering professor to discuss your ranking.

19. List ten skills or attributes that you need to work effectively with other people. How can you go about acquiring these skills and attributes?

20. Find out if your engineering college has a list of attributes it strives to impart to its graduates. How does it compare with the list in Section 1.4?

21. Rate yourself on a scale of 0 to ten (ten being highest) on the following items:

DESCRIPTION	RATING
Writing skills	
Oral communication skills	
Ability to work on teams	
Commitment to becoming an engineer	
Understanding of professional and ethical responsibility	
Recognition of the need for life-long learning	
Knowledge of contemporary issues	
Computer skills	

Ability to apply knowledge of mathematics	
Ability to apply knowledge of science	
Participation in student organizations	
Degree you work collaboratively with other students	
Time and energy devoted to studying	
Time spent on campus	
Overall grade point average	

22. Rate the items in Problem 21 above on a scale of 0 to ten (ten being highest) as to their importance.

23. Develop a method for determining which of the items in Problem 21 need your greatest attention (Hint: Use the 2x2 matrix below). Which quadrant contains items that need your greatest attention? Which quadrant contains items that need the least attention?

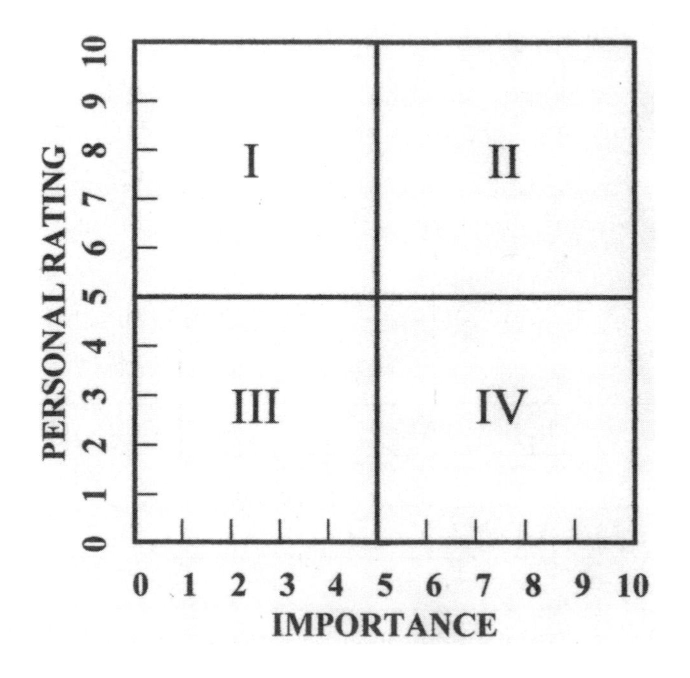

24. From the list in Problem 21, pick the three items that need your greatest attention and the three items that need your least attention. Develop a plan for self-improvement for those that need your greatest attention. Implement the plan.

Need my greatest attention	Need my least attention

25. Which of the items in Problem 21 have to do with your skills? With your attitude? With your approach to your studies?

26. Make a list of factors that are interfering with your ability to perform academically up to your full potential. How many of these are external to you (e.g., job, family, friends)? How many are internal (e.g., lack of motivation, poor study habits, etc.)? Which of these interferences can you reduce or eliminate completely? Develop a plan to do so.

27. Apply the "60-Hour Rule" presented in Section 1.5 to your situation. Based on that rule, how many credit hours should you be taking? How many are you taking? Are you overcommitted? What can you do about it?

28. Who are your best friends? Are they engineering majors? How many engineering majors do you know by name? What percentage of the students in your key math, science, and engineering classes do you know? How could you get to know more of them?

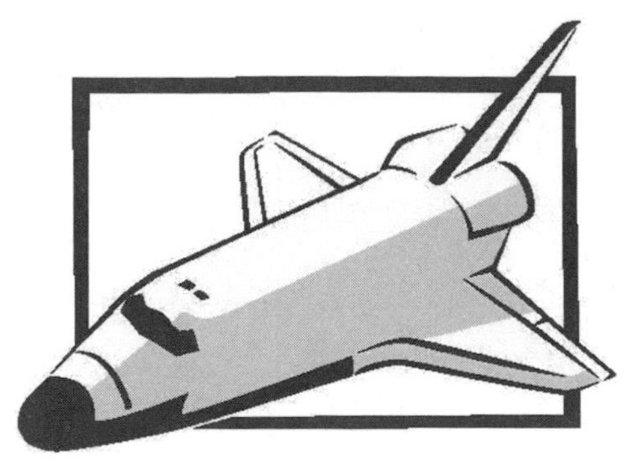

CHAPTER 2
The Engineering Profession

Scientists investigate that which already is;
Engineers create that which never has been.

Theodore Von Karman

INTRODUCTION

This chapter will introduce you to the engineering profession. Look at it as a discussion of "everything you ever wanted to know about engineering"—and then some. Hopefully, when you are finished reading the chapter, you will have a comprehensive understanding of the engineering profession and perhaps have found the engineering niche that attracts you most. This information, coupled with knowledge of the personal benefits you will reap from the profession, is intended to strengthen your commitment to completing your engineering degree.

First, we'll answer the question, "What is engineering?" Through several standard definitions, you'll learn that engineering is essentially the application of mathematics and science to develop useful products or processes. We'll then discuss the engineering design process, which we will demonstrate through a case study of an actual student design project.

To expand your understanding of engineering, we will take stock of the Greatest Engineering Achievements of the 20th Century, selected by the National Academy of Engineering and announced during National Engineers Week 2000. These achievements will show you the critical role engineering plays in making the quality of our life possible.

Next, we will discuss the rewards and opportunities that will come to you when achieve your B.S. degree in engineering. Having a clear picture of the many payoffs will be a key factor in motivating you to make the personal choices and put forth the effort required to succeed in such a challenging and demanding field of study.

We will then examine the various engineering disciplines, the job functions performed by engineers, and the major industry sectors that employ engineers. At the same time, we will open your horizons to the future by describing those fields showing the greatest promise for growth.

The last section of the chapter will focus on engineering as a profession, including the role of professional societies and the importance of professional registration.

2.1 What Is Engineering?

I'm sure you have been asked, ***"What is engineering?"*** I remember my grandmother asking me that question when I was in college. At the time, I didn't have much of an answer. Yet, when you think about it, it is a fundamental question, especially for new engineering students like yourself. So, just what is engineering?

A good starting point for answering this question is the theme for many years of *National Engineers Week*, held each February in honor of George Washington, considered our nation's first engineer. That theme depicts engineering according to its function:

Turning ideas into reality

To reflect the global nature of engineering, starting in 2006, the theme of National Engineers Week was changed to:

Engineers make a world of difference

Both themes are helpful in understanding engineering. The earlier theme suggests what engineers do. The more recent theme underscores the impact what engineers do has on the world.

Over the years, many definitions of engineering have been put forth, from that of the famous scientist Count Rumford over 200 years ago:

Engineering is the application of science to the common purpose of life

to the current standard definition of engineering provided by the Accreditation Board for Engineering and Technology (ABET) [1]:

> "Engineering is the profession in which a knowledge of the mathematical and natural sciences, gained by study, experience, and practice, is applied with judgment to develop ways to utilize, economically, the materials and forces of nature for the benefit of [hu]mankind. "

Harry T. Roman, a well-known New Jersey inventor and electrical engineer, compiled 21 notable definitions of engineering. These are presented in Appendix A.

EXER CISE

Study the 21 definitions of engineering presented in Appendix A. Then compose your own definition of engineering. Write it down and commit it to memory. This may seem like an unnecessary exercise, but I assure you it isn't. Aside from impressing others with a quick informed answer to the question "What is engineering?" this exercise will help clarify your personal understanding of the field.

ONE LAST POINT. The question is often asked: How is engineering different from science? An excellent answer was provided by Astronaut Neil Armstrong in the forward of *A Century of Innovation: Twenty Engineering Achievements That Changed Our Lives* [2]:

> ***Engineering*** is often associated with science and understandably so. Both make extensive use of mathematics, and engineering requires a solid scientific basis. Yet as any scientist or engineer will tell you, they are quite different. Science is a quest for "truth for its own sake," for an ever more exact understanding of the natural world. It explains the change in the viscosity of a liquid as its temperature is varied, the release of heat when water vapor condenses, and the reproductive process of plants. It determines the speed of light. Engineering turns those explanations and understandings into new or improved machines, technologies, and processes—to bring reality to ideas and to provide solutions to societal needs.

LEARNING MORE ABOUT ENGINEERING

As you learn more about the field of engineering, you will find there is no simple answer to the question, "What is engineering?" Because engineers do so many different things and perform so many different functions, learning about engineering is a lifelong process. Still, there are a variety of ways to start this process of learning about and understanding engineering, one being the tremendous amount of information you can access through the Internet.

One helpful web site you should check out is the one connected to National Engineers Week:

http://www.eweek.org

At that web site you can learn much about both engineering and National Engineers Week at the same time. And while at the web site, click on "New Faces in Engineering" and you'll have the opportunity to check out the accomplishments of young engineers (two to five years out of school) who are doing particularly interesting and unique work. Their e-mail addresses are included and you are invited to send your questions to them.

Additional web sites published by other professional engineering societies and the Federal government, such as those listed below, will help further your understanding of the field. The addresses of these sites are:

http://www.engineeringk12.org
http://www.discoverengineering.org
http://www.careercornerstone.org
http://www.jets.org
http://dedicatedengineers.org
http://www.bls.gov/oco/ocos027.htm

I would encourage you to research these web sites to broaden your understanding of engineering. As indicated in Chapter 1, Section 1.4, increasing your knowledge about engineering will strengthen your commitment to engineering study and your motivation to succeed.

2.2 The Engineering Process

At the heart of engineering is the *engineering process*, sometimes called the *engineering design process*. The engineering design process is a step-by-step method to produce a device, structure, or system that satisfies a need.

Sometimes this need comes from an external source. For example, the U.S. Air Force might need a missile system to launch a 1,000-pound communications satellite into synchronous orbit around the earth. Other times, the need arises from ideas identified within a company. For example, consumers did not initiate the need for various sizes of little rectangular yellow papers that would stick onto almost anything yet be removed easily when 3M invented "Post-its" [3].

Whatever the source, the need is generally translated into a set of specifications ("specs"). These can include performance specifications (e.g., weight, size, speed, safety, reliability), economic specifications (e.g., cost), and scheduling specifications (e.g., production and delivery dates).

YOUR ALARM CLOCK IS AN EXAMPLE

Virtually everything around you was designed by engineers to meet certain specifications. Take the start of your day, for example. You probably wake up to a battery-powered alarm clock. Every design feature of the clock was carefully considered to meet detailed specifications. The alarm was designed to be loud enough to wake you up but not so loud as to startle you. It may even have a feature in which the sound level starts very low and increases progressively until you wake up. The digital display on your clock was designed to be visible day and night. The batteries were designed to meet life, safety, and reliability requirements. Economic considerations dictated material selection and manufacturing processes. The clock also had to look aesthetically pleasing to attract customers, while maintaining its structural integrity under impact loading, such as falling off your night stand.

THE ENGINEERING DESIGN PROCESS

Now that you have been introduced to the first two steps—identifying the need and then drawing up specifications to meet that need—the complete step-by-step design process can be illustrated by the schematic below.

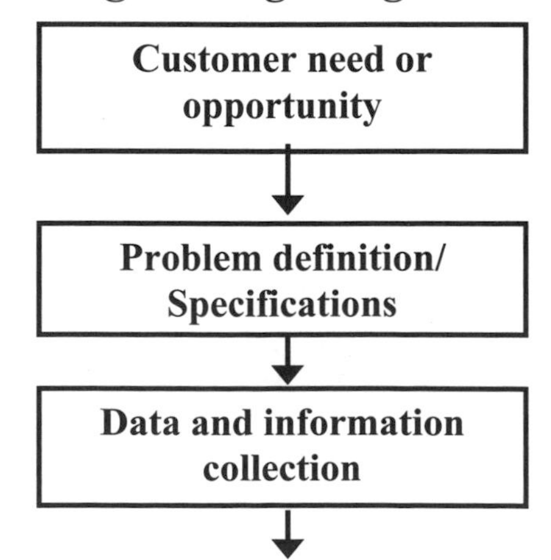

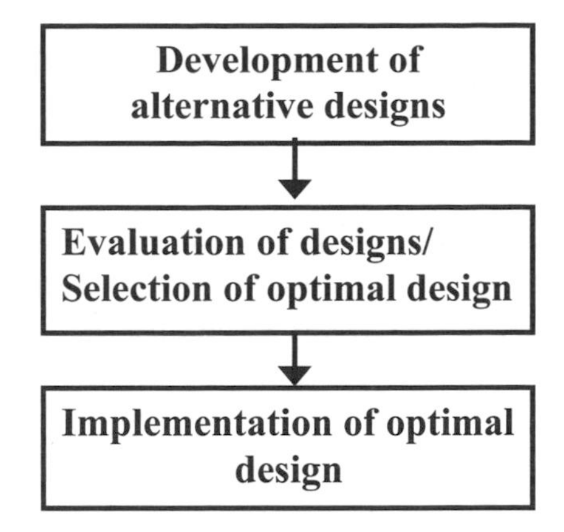

From this schematic, you can see that each step of the design process reflects a very logical, thorough problem-solving process. The problem definition and specifications (Steps 1 and 2) will need to be supplemented by additional data and information (Step 3) before the development of possible solutions can begin (Step 4). The process of developing and evaluating possible designs (Steps 4 and 5) involves not only creativity but also the use of computer-aided drafting (CAD), stress analysis, computer modeling, material science, and manufacturing processes. Engineers also bring a great deal of common sense and experience to the design process.

During the design process, a number of constraints may be identified. Whatever these constraints may be—e.g., availability of parts and materials, personnel, and/or facilities—the final design must not only meet all design specifications but also satisfy any constraints.

Many iterations through the engineering design process may be required before a final design is selected. Fabrication of some of the designs may be required in order to test how well each meets the performance specifications.

It should be noted that the engineering design process is part of the broader product development cycle that begins with the perception of a market opportunity and ends with the production, sale, and delivery of a product. An excellent resource on this subject is the book *Product Design and Development* by Karl Ulrich and Steven Eppinger [4].

2.3 Case Study: Solar-Powered Electric Vehicle

The six steps of the engineering design process make most sense when they are seen in action. We are using an actual case study about the design and construction of a solar-powered vehicle so you can see each step of the process at work.

CUSTOMER NEED OR OPPORTUNITY

In April, 2004, Delft University of Technology in the Netherlands received an invitation from the South Australian Tourism Commission to apply for entry into the 2005 World Solar Challenge. This would be the 8th time since the inception of the race in 1987 that solar-electric vehicles built by teams of engineers at both universities and corporations would race 3,000 km across the Australian continent from Darwin to Adelaide. In this case, the opportunity or need—the first step in the engineering design process—was created by the South Australian Tourism Commission. For information about the World Solar Challenge go to the official web site at:

http://www.wsc.org.au

It was an easy decision for the Delft University team since they were the defending champions, having won both the 2001 and 2003 World Solar Challenge events with their Nuna and Nuna 2 vehicles, respectively. The decision was made to put together an eleven-member Nuna 3 student team to seek an unprecedented "three-peat."

PROBLEM DEFINITION AND SPECIFICATIONS

The primary design specifications, Step 2 of the engineering design process, were established by the race rules. They included the following requirements:

- Maximum vehicle size: Length = 5 meters; Width = 1.8 meters; Height = 1.6 meters
- Minimum height for driver's eyes: 700 mm
- Solar cell type: No limitation
- Battery type: Commercially available
- Maximum battery capacity: 5.5 Kw-hr
- Safety requirements: Safety belts, helmets, structural roll bar, 15-second unassisted driver egress, brakes, tires, driver vision, steering, electrical systems

For the Delft University Nuna 3 team, these requirements led to additional problem definitions and specifications. Who would lead the team? How much money would the entire project cost? How would it be financed? What facilities would be required? The race rules had specified size and height requirements, but what would be the optimal weight of the vehicle? What materials would be needed to fabricate the vehicle?

DATA AND INFORMATION COLLECTION

Before developing alternative designs that met all the design specifications, the team first had to collect extensive data and information. They needed to learn the technologies associated with electric motor systems, batteries, solar power systems, vehicle aerodynamics, the design and construction of light-weight structures, vehicle suspension and steering systems, mechanical drive systems, and wheels. They also needed to learn about the topography of the race route, expected weather conditions, and solar insolation estimates.

Much of this knowledge built on what had been learned through the Nuna and Nuna 2 projects. However, the challenge of designing an improved vehicle required greater attention to each design issue.

DEVELOPMENT OF ALTERNATIVE DESIGNS

Once they had collected sufficient basic data, the team moved to the next step of the design process: developing alternative designs.

Producing an optimally designed solar-electric vehicle is an excellent example of the type of design tradeoffs that often must be made during this stage of the design process. The Nuna 3 team realized that high performance could be achieved by the following:

- High solar pancl power
- Low aerodynamic drag (low drag coefficient; low frontal area)
- Low vehicle weight
- High electrical power system efficiency
- High mechanical drive system efficiency
- Good battery performance
- High overall reliability

However, several of these design parameters conflict. Achieving all of them simultaneously just isn't possible. This is where design tradeoffs enter into the development of alternative designs.

For the Nuna 3 vehicle, the need for certain tradeoffs was immediately apparent. For example, the team knew they could achieve high solar panel power through a large solar panel surface area. They also knew, however, that a large surface area would result in high drag and high vehicle weight.

Low vehicle weight was imperative, but at some point would contribute to poor structural integrity and low overall reliability. Another tradeoff involved the distribution of heavy electrical components (batteries, controllers, etc). Placing them close together would reduce wiring requirements thereby reducing weight and electrical losses. Distributing them optimally would improve vehicle handling and stability. Selection of solar cells required trading off panel power, cost, assembly time, and reliability.

An important step in this stage of the engineering design process is to select performance specifications or design targets for the key design parameters. For example, the Nuna 3 team felt that a top place finish in 2005 World Solar Challenge would be ensured if their vehicle performed to the following specifications:

- Solar panel power (peak) – 2,100 watts
- Aerodynamic drag (Cd x A) - 0.07
- Vehicle weight (exclusive of driver and batteries) – <200 kg
- Electric power system efficiency - >97%
- Battery weight – 30 kg (for 5.5 kw-hr capacity)

And the Nuna 3 team had more than winning in mind. The Nuna 2 vehicle had averaged 97 km/hr in winning the 2003 World Solar Challenge. No solar-powered car had ever topped the 100 km/hr mark in the previous seven races. The Nuna 3 team set its sights on this elusive target.

EVALUATION OF DESIGNS/SELECTION OF OPTIMAL DESIGN

This is one of the most difficult, challenging, and time-consuming steps in the engineering design process. For many engineers, however, it is also the most interesting and rewarding one, for here is where ideas really begin to turn into reality.

For the Nuna 3 team, this step was no different. In evaluating potential designs and selecting the optimal one, they still had numerous hurdles to overcome, and questions to resolve. Although they had faced many of these quandaries in the earlier stages of the design process, they now needed hard answers to such questions as:

- What should the external shape of the vehicle be?
- How can driver comfort be assured without adding to vehicle weight and frontal area?

- What composite materials should be used on specific areas of the vehicle? What fabrication process should be used?
- What fastening techniques (bolting, welding, gluing, riveting) should be used in a multitude of cases?
- What solar cells should be selected?
- What should be the design voltage of the solar panel?
- What motor and motor controller should be used?
- What type of tires and rims should be used?
- What type of brakes should be used?
- What type and how many batteries should be used?
- Which components could be fabricated by the team in-house and which would need to be contracted out?

A number of key questions were answered as follows:

SOLAR CELLS. Three-junction, Gallium Arsenide (GaAs) solar cells having an efficiency of over 26 percent were selected. Since the race was to be conducted in September, a time when the sun is lower in the sky, it was decided to place as many cells as possible on the sides of the vehicle.

AERODYNAMIC DESIGN. An ideal aerodynamic design was developed to minimize frontal area and overall drag coefficient through computer simulations of proposed designs, testing scale models in a small wind tunnel, and finally testing the full scale car in a large wind tunnel.

ELECTRIC MOTOR. The decision was made to refurbish the Drivetek motor used in the Nuna and Nuna 2 vehicles rather than purchase a new slightly more efficient motor. The "wheel motor" was totally encased in the rear wheel to minimize loss through mechanical transmission from the motor to the wheel. Improvements included a 30 percent reduction in weight and higher efficiency (>98%) at the peak power and torque levels.

BATTERIES. The team selected Lithium-polymer batteries made by the South Korean company Kokam. The battery pack considered of six modules of seven cells each, all connected in series. The batteries, the most energy-dense available, had a capacity of 5.5 kw-hr and a total weight of 30 kilograms.

After definitively answering these and other questions, the team settled on their optimal design. A series of drawings of all the parts followed, and the team advanced to the final step of the engineering design process: implementing the optimal design.

Implementation of Optimal Design

Now began the "real" work, as the title of this last phase of the design process indicates. The Nuna 3 team divided this part of the project into three stages.

The first stage consisted of building the mechanical system, including the overall structure and external body, wheels, steering, and brakes. Once this stage was complete, the vehicle could be pushed around a parking lot or rolled down a hill.

In stage two, they installed the power electronic system—including the motor, motor controls, batteries, and drive system. With this stage finished, the vehicle could be driven around as an actual electric vehicle.

In the third stage, the team installed the solar panels on the body. The panels were prefabricated by Gochermann Solar Technology in Germany according to design details provided by the Nuna 3 team. Needless to say, all of this work required extreme attention to detail, particularly in maintaining the smooth aerodynamic shape of the vehicle.

Nuna 3 Team Wins 2005 World Solar Challenge

Once the entire design process was completed, on June 21, 2005, the team proudly presented *Nuna 3* at a major media event held at the Circuit Park Zandvoort, a motor speedway near the Dutch beach resort of Zandvoort. The excitement of the entire nation was reflected by the fact that the Dutch band Sonar11 wrote a song "Gone with the Sun" especially for Nuna 3. The team's job was far from over, however. Lots of work remained, such as testing the vehicle's performance, formulating the race strategy, and transporting both the team and vehicle to the World Solar Challenge starting line in Darwin, Australia—a distance of more than 13,000 miles.

Soon the team and Nuna 3 vehicle were in Darwin for several weeks of pre-race activity including test driving and completion of a rigorous scrutineering process. This was the first chance to see the other members of the 21-car field. Front runners included: *Aurora* from Australia, a past winner of the World Solar Challenge; *Momentum* from the University of Michigan, winner of the 2005 North American Solar Challenge; *Tesserac*t from MIT, always a top finisher in solar car racing; and *Sky Ace Tiga* from Ashiya University in Japan, holder of the world speed record for solar powered cars of 165 km/hr.

On the fourth day, the Nuna 3 crossed the finish line at Adelaide, Australia, 3 hours and 24 minutes ahead of second place Aurora. The team accomplished its goal of exceeding an average speed of 100 km/hr, covering the 3,000 kilometer distance in a record 29 hours and 11 minutes, averaging 103 km/hr. For more information on the Nuna 3 project see:

http://www.nuonsolarteam.nl

More information about solar car racing and the design of solar cars can be found in References 5, 6, and 7.

Nuna 3 Solar Powered Vehicle During Test Driving

(Photo by Hans-Peter van Velthoven/Nuon)

THE NEEDS AND OPPORTUNITIES FOR ENGINEERING DESIGN ARE BOUNDLESS

The purpose of chronicling Delft University's solar car project was to illustrate the engineering design process in action. Now that you have seen the logic and demand that each step of the process entails, you should easily be able to come up with a list of the many other problems, needs, and opportunities that would suit its step-by-step approach. Here are just a few ideas that occurred to me. What ideas would you add to this list? Remember, it is entirely possible that, down the road, you will be the engineer who turns one of these needs into reality.

- A device carried by a police officer that would detect a bullet fired at the officer and intercept it.

- A device that would mark the precise location of a football when the referee blows the whistle.
- A system that would not permit an automobile to be stolen.
- A device that would program a VCR to skip the commercials while taping your favorite TV show.
- A machine that would serve ping-pong balls at different speeds and with different spins.
- A device that identifies vehicles that are carrying explosives.
- A car alarm that goes off if the driver falls asleep.
- A device that cuts copper tubing in tight places.
- An in-home composting and recycling system that eliminates the need for sewer or septic systems.
- A device that prevents elderly people from being injured when they fall down.
- An affordable, fuel-cell powered automobile that only emits water vapor.
- A system that continues to tape your favorite morning radio show after you arrive at work so you can listen to it on the way home.
- A high-rise building with an "active suspension system" that responds to ground movement (earthquakes).

2.4 Greatest Engineering Achievements of the 20th Century

Although engineering achievements have contributed to the quality of human life for more than 5,000 years [8], the 20th century stands out for its remarkable engineering progress and innovation. In recognition of this, as we entered the 21st century the National Academy of Engineering (NAE) launched a project to select the 20 "Greatest Engineering Achievements of the 20th Century."

The primary selection criterion was the impact of the engineering achievement on the quality of life in the 20th century. William A. Wulf, president of the National Academy of Engineering summed it up well:

> "Engineering is all around us, so people often take it for granted, like air and water. Ask yourself, what do I touch that is not engineered? Engineering develops and delivers consumer goods, builds the networks of highways, air and rail travel, and the Internet, mass produces antibiotics, creates artificial heart valves, builds lasers, and offers such wonders as imaging technology and conveniences like microwave ovens and compact discs. In short, engineers make our quality of life possible."

The following is a listing of the "Greatest Engineering Achievements" presented by Neil Armstrong at the National Press Club in Washington, D.C. on February 22, 2000.

#20 - High Performance Materials
#19 - Nuclear Technologies
#18 - Laser and Fiber Optics
#17 - Petroleum and Gas Technologies
#16 - Health Technologies
#15 - Household Appliances
#14 - Imaging Technologies
#13 - Internet
#12 - Space Exploration
#11 - Interstate Highways
#10 - Air Conditioning and Refrigeration
#9 - Telephone
#8 - Computers
#7 - Agricultural Mechanization
#6 - Radio and Television
#5 - Electronics
#4 - Safe and Abundant Water
#3 - Airplane
#2 - Automobile
#1 - Electrification

Brief descriptions of each is presented in Appendix B. Detailed descriptions of each "great achievement" can be found in an excellent book titled *A Century of Innovation: Twenty Engineering Achievements that Transformed Our Lives* [2] and also on the web at:

http://www.greatachievements.org

2.5 Rewards and Opportunities of an Engineering Career

Engineering is a unique and highly selective profession. Among the 130 million people employed in the United States, only about 1.4 million (1.1 percent) list engineering as their primary occupation [9]. This means the overwhelming majority of people employed in this country do something **other than engineering**.

These employment figures are reflected by national college and university statistics. Engineering typically represents slightly less than five percent of college graduates, as the following table shows [10]:

Major	Number of 2003/04 College Graduates	Percent of Total
Business	307,149	21.9%
Social Sciences	150,357	10.7%
Education	106,278	7.6%
Math and Science	92,819	6.6%
Psychology	82,098	5.9%
Visual and Performing Arts	77,181	5.5%
Health Professions	73,934	5.2%
Engineering	**63,558**	**4.5%**
TOTAL	1,399,542	100.0%

So why choose to study engineering? Why strive to become one of those 4.5 percent of college graduates who receive their B.S. degree in engineering? I'll tell you why.

The benefits of an engineering education and the rewards and opportunities of a career in engineering are numerous. I have frequently led first-year engineering students in a brainstorming exercise to identify these many rewards and benefits. We generally develop a list of 30 to 40 items, which each student then ranks (or deletes) according to personal preferences. For one individual, being well paid may be #1. Someone else may be attracted by the opportunity to do challenging work. Still others may value engineering because it will enable them to make a difference in people's lives.

My personal top ten list is presented on the next page. Although your list may well differ from mine, I am going to discuss each briefly—if only to help you realize more fully the many rewards, benefits, and opportunities an engineering career holds for you.

Ray's Top Ten List

1. **Job Satisfaction**
2. **Varied Opportunities**
3. **Challenging Work**
4. **Intellectual Development**
5. **Social Impact**
6. **Financial Security**
7. **Prestige**
8. **Professional Environment**
9. **Understanding How Things Work**
10. **Creative Thinking**

After studying my list and developing your own, hopefully you will find yourself more determined to complete your engineering studies. You may also find yourself somewhat puzzled by the skewed statistics that opened this section. With so many benefits and job opportunities a career in engineering promises, you'd think that college students would be declaring engineering majors in droves.

I guess engineering really is a unique and highly selective profession. Consider yourself lucky to be one of the "chosen few."

1. JOB SATISFACTION

What would you say is the #1 cause of unhappiness among people in the United States? Health problems? Family problems? Financial problems? No. Studies have shown that, by far, the #1 cause of unhappiness among people in the U.S. is **job dissatisfaction**. And furthermore, Americans are growing increasingly unhappy with their jobs. A study conducted in 2005 for the Conference Board, a leading business membership and research organization, indicated that only half of all Americans are satisfied with their jobs, down from nearly 60 percent in 1995 [11].

Do you know people who dislike their job? People who get up every morning and wish they didn't have to go to work? People who watch the clock all day and can't wait until their workday is over? People who look forward to Fridays and dread Mondays? People who work only to earn an income so they can enjoy their time off? Maybe you have been in one of these situations. Lots of people are.

Throughout my career, it has been very important to me to enjoy my work. After all, I spend eight hours or more a day, five days a week, 50 weeks a year, for 30 or 40 years working. This represents about 40 percent of my waking time. Which would you prefer? Spending 40 percent of your life in a career (or series of jobs) you despise? Or spending that 40 percent in a career you enjoy and love? I'm sure you can see why it is extremely important to find a life's work that is satisfying, work that you want to do.

Engineering could very well be that life's work. It certainly has been for me and for many of my colleagues over the years. But what exactly does "job satisfaction" mean? The remaining items on my "Top Ten List" address this question. Remember, though, these are my preferences; yours may very well be different.

2. Varied Opportunities

While the major purpose of this chapter is to help you understand the engineering profession, you have just skimmed the surface thus far. Your introduction to the engineering field has largely been a "functional" one, starting with the idea that engineering is the process of "turning ideas into reality," followed by a detailed look at the engineering design process—more function.

As you'll learn subsequently, engineering entails much more than just "functions" governed by a rigid six-step design process. In fact, I like to think of engineering as a field that touches almost every aspect of a person's life. I often point out to students that the day you walk up the aisle to receive your B.S. degree in engineering, you have closed no doors. **There is nothing you cannot become from that time forward!** Doctor. Lawyer. Politician. Astronaut. Entrepreneur. Teacher. Manager. Salesperson. Practicing engineer. All these and many others career opportunities are possible.

Here are some examples of people educated as engineers and the professions they ended up in:

ENGINEER	PROFESSION
Neil Armstrong	Astronaut/First Person on Moon
Herbert Hoover	President of the United States
Jimmy Carter	President of the United States
Alfred Hitchcock	Movie Director
Eleanor Baum	Dean of Engineering
Herbie Hancock	Jazz Musician
Paul MacCready	Inventor (Designer GM EV1 Electric Car)
Ellen Ochoa	Space Shuttle Astronaut
Hyman G. Rickover	Father of the Nuclear Navy
Bill Nye	Host of TV Show "Bill Nye The Science Guy"
Boris Yeltsin	Former President of Russia
Alexander Calder	Sculptor
Bill Koch	Yachtsman (Captain of America Cup Team)
W. Edwards Deming	Father of Modern Management Practice
Grace Murray Hopper	U.S. Navy Rear Admiral/Computer Engineer
Ming Tsai	Restaurateur and Star of TV Cooking Show
Montel Williams	Syndicated Talk Show Host
Samuel Bodman	U.S. Secretary of Energy
Michael Bloomberg	Billionaire/Mayor of New York City
A. Scott Crossfield	X-15 Test Pilot
Don Louis A. Ferre'	Governor of Puerto Rico
Yasser Arafat	Palestinian Leader/Nobel Peace Prize Laureate
Tom Landry	Former Dallas Cowboy's Head Coach
Shiela Widnall	Former Secretary of the Air Force
Robert A. Moog	Father of Synthetic Music
Chester Carlson	Inventor of Xerox Process
John A. McCone	Director of Central Intelligence Agency
Arthur C. Nielsen	Developer of Nielsen TV Ratings

Although none of the above individuals ended up working as a practicing engineer, I expect they would all tell you that their engineering education was a key factor in their subsequent successes. You can learn more about these and other famous "engineers" at:

http://www.engineeringk12.org/students/What_Is_Engineering/default.php
(Click on "Famous Engineers")

Personal Story

When I was an engineering student, I had no idea that the career path I have taken even existed. After completing my B.S. and M.S. degrees in Mechanical Engineering at MIT, I worked for five years as a practicing engineer at Rocketdyne, a division of Rockwell International at that time. While doing some part-time teaching to supplement my salary, I developed an interest in an academic career and was able to get a position on the engineering faculty at California State University, Northridge.

Although I enjoyed teaching, my interests shifted more to administration and working with students outside of the classroom. I started the first Minority Engineering Program in California and directed it for ten years. The administrative and management experience I gained led me to the position of Dean of Engineering. My engineering career thus evolved from practicing engineering to teaching it; from teaching it to creating and directing a special program for minority engineering students; and finally from directing a program to managing an entire engineering college.

The field of engineering practice itself offers an enormous diversity of job functions. There are analytical engineers, design engineers, test engineers, development engineers, sales engineers, and field service engineers. The work of analytical engineers most closely resembles the mathematical modeling of physical problems you do in school. But only about ten percent of all engineers fall into this category, pointing to the fact that engineering *study* and engineering *work* can be quite different.

- If you are imaginative and creative, **design engineering** may be for you.
- If you like working in laboratories and conducting experiments, you might consider **test engineering**.

- If you like to organize and expedite projects, look into becoming a **development engineer**.
- If you are persuasive and like working with people, **sales or field service engineering** may be for you.

Later in this chapter, we will examine the wide variety of engineering job functions in more detail. Then, in Chapter 8, we will explore less traditional career paths for which engineering study is excellent preparation, such as medicine, law, and business.

3. Challenging Work

Do you like intellectual stimulation? Do you enjoy tackling challenging problems? If so, you'll get plenty of both in engineering. Certainly, during your period as an engineering student, you will face many challenging problems. But, as the saying goes, "you ain't seen nothing yet" until you graduate and enter the engineering work world, where there is no shortage of challenging, "open-ended" problems. By "open-ended," I mean there is generally no one "correct" solution, unlike the problems you usually are assigned in school. Open-ended problems typically generate many possible solutions, all of which equally meet the required specifications. Your job is to select the "best" one of these and then convince others that your choice is indeed the optimal one.

It certainly would be helpful if you had more exposure to open-ended problems in school. But such problems are difficult for professors create, take more time for students to solve, and are excessively time-consuming to grade. Regardless, however, of the kind of problem you are assigned (open-ended or single answer; in school or the engineering work-world), they all challenge your knowledge, creativity, and problem-solving skills. If such challenges appeal to you, then engineering could be a very rewarding career.

4. Intellectual Development

Engineering education "exercises" your brain much the way weight-lifting or aerobics exercises your body—and the results are remarkably similar. The only difference is that physical exercise improves your body, while mental exercise improves your mind. As your engineering studies progress, therefore, your abilities to solve problems and think critically will increasingly grow stronger.

This connection between mental exercise and growth is by no means "news" to educators. But recent research in the cognitive sciences has uncovered knowledge that explains how and why this process works [12]. We now know, for example, that the brain is made up of as many as 180 billion neuron cells. Each neuron has a very large number of tentacle-like protrusions called dendrites. The dendrites make it possible for each neuron to receive signals (synapses) from thousands of neighboring neurons. The extent of these "neural networks" is determined in large part by the demands we place on our brains—i.e., the "calisthenics" we require of them. So the next time your find yourself reluctant to do a homework assignment or study for a test, just think of all those neural networks you could be building.

One of the things I value most about my engineering education is that it has developed my logical thinking ability. I have a great deal of confidence in my ability to deal effectively with problems. And this is not limited to engineering problems. I am able to use the critical thinking and problem-solving skills I developed through my engineering education to take on such varied tasks as planning a vacation, searching for a job, dealing with my car breaking down in the desert, organizing a banquet to raise money, purchasing a new home, or writing this book. I'm sure you also will come to value the role your engineering education plays in your intellectual growth.

5. Social Impact

I hope you are motivated by a need to do something worthwhile in your career, something to benefit society. Engineering can certainly be an excellent career choice to fulfill such humanitarian goals.

The truth is, just about everything engineers do benefits society in some way. Engineers develop transportation systems that help people and products move about so easily. Engineers design the buildings we live and work in. Engineers devise the systems that deliver our water and electricity, design the machinery that produces our food, and develop the medical equipment that keeps us healthy. Almost everything we use was made possible by engineers.

Depending on your value system, you may not view all engineering work as benefiting people. Some engineers, for example, design military equipment like missiles, tanks, bombs, artillery, and fighter airplanes. Others are involved in the production of pesticides, cigarettes, liquor, fluorocarbons, and asbestos. As an engineer, you will need to weigh the merits of such engineering functions and make your career choices accordingly.

My view is that engineering holds many more beneficial outcomes for society than detrimental ones. For example, opportunities exist for engineers to use their expertise in projects designed to clean up the environment, develop prosthetic aids, develop clean and efficient transportation systems, find new sources of energy, solve the world's hunger problems, and improve the standard of living in underdeveloped countries.

6. Financial Security

When I ask a class of students to list the rewards and opportunities that success in engineering study will bring them, money is almost always #1. In my "Top Ten List," it's #6. It's not that engineers don't make good money. They do! It's just that money is not a primary motivator of mine.

I've always held the view that if you choose something you like doing, work hard at it, and do it well, the money will take care of itself. In my case, it has. Of course, you may discount my philosophy because of my credentials and career successes. But remember, my engineering career began much the same way yours will—working in industry as a practicing engineer. My subsequent career moves, however, were never motivated by money alone. I hope you too don't make money your primary reason for becoming an engineer. Other reasons, like job satisfaction, challenging work, intellectual development, and opportunities to benefit society hopefully will prove to be more important factors. If they are, you will find the quality of your life enriched tremendously. And I guarantee that "the money will take care of itself," as it has for me.

Let's not lose sight of reality, however! If you do become an engineer, **you will be rewarded financially**. Engineers, even in entry-level positions, are well paid. In fact, engineering graduates receive the highest starting salary of any discipline, as shown in the data below for 2005/06 [13].

Beginning Offers to 2005/06 Graduates

Discipline	Avg. Salary
Engineering	**$51,465**
Computer Sciences	49,680
Engineering Technology	48,514
Nursing	45,347

Business	41,900
Mathematics and Sciences	38,217
Agriculture & Natural Resources	33,716
Education	32,438
Humanities & Social Sciences	31,290
Communications	31,110

You also may be interested to know that of the 20,491 offers reported in this study, 7,964 (38.9 percent) went to business graduates and 5,160 (25.1 percent) went to engineering graduates—disciplines that comprise only 26 percent of all college graduates. The remaining 74 percent received only 36 percent of all job offers. Put another way:

Engineering graduates received almost six times as many job offers as the average number for graduates in all other disciplines.

If the starting salary data has not convinced you that engineering is a financially rewarding career, perhaps you will be convinced by the fact that many of the world's wealthiest people started their careers with a degree in engineering. You will find a listing of some of these people in Appendix C. As reported by *Forbes Magazine* [14], the personal wealth of these individuals ranges from a high of $30 billion down to $4 billion. I hope this brings the idea home that "the sky is the limit."

EXERCISE

Pick one of the 21 individuals listed in Appendix C among the world's wealthiest *engineers*. Find out as much as you can about the person by conducting an Internet search using a search engine such as Google. What was their major in college? What did they do early in their career? How did they become so wealthy? What lessons can you learn from their success?

7. PRESTIGE

What is *prestige*? The dictionary defines it as "the power to command admiration or esteem," usually derived from one's social status, achievements, or profession. Engineering, as both a field of study and a profession, confers prestige. You may have already experienced the prestige associated with being an engineering major. Perhaps you have stopped on campus to talk with another student and during the conversation, he or she asked, "What's your major?" What reaction did you get when you said, "Engineering"? Probably one of respect, awe, or even envy. To non-engineering majors, engineering students are "the really smart, studious ones." Then, if you reciprocated by asking about that student's major, you may wish you hadn't after getting an apologetic response like, "I'm still undecided."

This hypothetical conversation between an engineering and non-engineering student is not farfetched. In fact, variations of it take place all the time. Everyone knows that engineering study requires hard work, so people assume you must be a serious, highly capable student.

I often ask students to name a profession that is more prestigious than engineering. "Medicine" always comes up first. I tend to agree. Physicians are well paid and highly respected for their knowledge and commitment to helping people live healthier lives. So if you think you want to be a medical doctor and have the ability, arrange to meet with a pre-med advisor as soon as possible and get started on your program. I certainly want to havc the most capable people as my doctors.

After medicine, law and accounting are typically cited as more prestigious professions than engineering. Here, however, I disagree, arguing against these and **every other profession** as conferring more prestige than engineering. Anyone who knows anything about engineering would agree that engineers play critical, ubiquitous roles in sustaining our nation's international competitiveness, in maintaining our standard of living, in ensuring a strong national security, in improving our health, and in protecting public safety. I can't think of any other profession that affects our lives in so many vital, significant ways.

Engineers are critical to our:

8. Professional Environment

Although engineers can perform a variety of functions and work in many different settings, most new engineering graduates are hired into entry-level positions in "hi-tech" companies. While the nature of your work and status within the company may quickly change, there are certain standard characteristics of all professional engineering work environments.

For one, you will be treated with respect—both by your engineering colleagues and by other professionals. With this respect will come a certain amount of freedom in choosing your work and, increasingly, you will be in a position to influence the directions taken by your organization.

As a professional, you also will be provided with adequate workspace, along with whatever equipment and staff support you need to get your work done.

Another feature of the engineering work environment is the many opportunities you will have to enhance your knowledge, skills, self-confidence, and overall ethos as a professional engineer. Experienced engineers and managers know that new engineering graduates need help in making the transition from college to the "real world." From the outset, then, your immediate supervisor will closely mentor you, giving you the time and guidance to make you feel "at home" in your new environment. She will carefully oversee your work assignments, giving you progressively more challenging tasks and teaming you with experienced engineers who will teach you about engineering and the corporate world.

Once you are acclimated to your new position, your company will see to it that your engineering education and professional development continue. You will frequently be sent to seminars and short courses on a variety of topics, from new engineering methods to interpersonal communications. You may be given a travel allotment so you can attend national conferences of professional engineering societies. You also may discover that your company has an educational reimbursement program that will pay your tuition and fees to take courses at a local university for professional development or to pursue a graduate degree.

You can expect yearly formal assessments of your performance, judged on the merits of your contributions to the company. As a professional, you will not be required to punch a clock, for your superiors will be more concerned about the quantity and quality of your work, not your "time-on-tasks." If you have performed well in these areas, you can usually expect an annual merit salary increase, plus occasional bonuses for a "job particularly well done." Promotions to higher positions are another possibility, although they generally have to be earned over an extended period of time.

Finally, as a professional, you will receive liberal benefits, which typically include a retirement plan, life insurance, medical insurance, dental insurance, sick leave, paid vacation and holidays, and savings or profit-sharing plans.

9. UNDERSTANDING HOW THINGS WORK

Do you know why golf balls have dimples on them? Do you understand how the loads are transmitted to the supports on a suspension bridge? Do you know what nanotechnology is? How optical storage devices work? How fuel cells work? When you drive on a mountain road, do you look at the guard rails and understand why they were designed the way they are? Do you know why split-level houses experience more damage in earthquakes? Do you know why we use alternating current (AC) rather than direct current (DC)? One of the most valuable outcomes of my engineering education is understanding how things around me work.

Furthermore, there are many issues facing our society that depend on an understanding of technology. Why don't we have more zero-emission electric vehicles rather than highly polluting cars powered by internal combustion engines? Should we have stopped building nuclear reactors? What will we use for energy when the earth's supply of oil becomes prohibitively costly or runs out? Can we count on nuclear fusion? Should we have supersonic aircraft, high-speed trains, and automated highways? Is it technically feasible to develop a "Star Wars" defense system that will protect us against nuclear attack? Why are the Japanese building higher quality automobiles than we are building? Can we produce enough food to eliminate world hunger? Do high-voltage power lines cause cancer in people who live or play near them?

Your engineering education will equip you to understand the world around you and to develop informed views regarding important social, political, and economic issues facing our nation and the world. Who knows? Maybe this understanding will lead you into politics.

10. CREATIVE THINKING

Engineering is by its very nature a creative profession. The word "engineer" comes from the same Latin word *ingenium* as the words "genius" and "ingenious." This etymological connection is no accident: engineers have limitless opportunities to be ingenious, inventive, and creative. Do you remember reading about the "Greatest Engineering Achievements of the 20th Century"? You can be sure that creativity played a major role in each of these achievements.

Sometimes new engineering students have difficulty linking "creativity" with "engineering." That's because, at first glance, the terms are likely to invoke their stereotypical connections: "creativity" with art; "engineering" with math, science, and problem solving. The truth, though, is that creativity is practically an essential ingredient of engineering. Consider, for example, the following definition of "creativity," taken from a book entitled *Creative Problem Solving and Engineering Design* [15]:

> ***Playing with imagination and possibilities while interacting with ideas, people, and the environment, thus leading to new and meaningful connections and outcomes.***

This is just what engineers do. In fact, this definition of "creativity" could almost be a definition of "engineering."

To experienced engineers, who regularly engage in solving open-ended, real-world problems, the need for creativity in the engineering process is a given. It would seem particularly important, for example, during Steps 4, 5, and 6 of the engineering design process described in Section 2.2, which involve developing and evaluating alternative possible solutions, followed by the selection of the "best" one. Without an injection of creativity in these steps, the actual "best" solution may be overlooked entirely.

However, these are not the only steps of the engineering design process that involve creativity. Indeed, creativity enters into <u>every</u> step of the process. It would be a good exercise for you to review the six steps of the engineering design process to see how creativity can come into play at each step.

Beyond the engineering process itself, the need for engineers to think creatively is greater now than ever before, because we are in a time when the rate of social and technological changes has greatly accelerated. Only through creativity can we cope with and adapt to these changes. If you like to question, explore, invent, discover, and create, then engineering would be an ideal profession for you.

A wonderful place to explore the way human creativity in art, technology, and ideas has shaped our culture is ***The Engines of Our Ingenuity*** web page:

http://www.uh.edu/engines

There you will find the text of more than 2,150 episodes created and presented on National Public Radio for almost two decades by John Lienhard, professor of mechanical engineering at the University of Houston.

REFLECTION

Review my top ten list of "rewards and opportunities" that will come to you if you are successful in getting your engineering degree. Which one on the list is the most important to you? Money? Prestige? Challenging work? Making a difference in the world? Reflect on the one you chose. Why did you choose it? Why is that one important to you?

2.6 ENGINEERING DISCIPLINES

At this point you should have a general understanding of what engineering is and what engineers do—along, of course, with the many rewards and opportunities that engineering offers. Our goal in the remainder of this chapter is to clarify and broaden that understanding. We'll start by looking at engineering from a new perspective, and that is how engineers can be classified by their academic discipline.

Until recently, engineering has consisted of five major disciplines, which enroll the largest number of students. In rank order, these disciplines are:

- **Mechanical Engineering**
- **Electrical Engineering**
- **Civil Engineering**
- **Chemical Engineering**
- **Industrial Engineering**

A sixth discipline, **Computer Engineering**, has now been added to this list. Initially a subspecialty within electrical engineering (and still organized that way at many institutions), computer engineering has grown so rapidly that universities are increasingly offering separate accredited B.S. degrees in this field. (Given these changes, computer engineering is treated separately in my subsequent discussion of engineering disciplines.)

In addition to the top six disciplines, there are many other more specialized, non-traditional fields of engineering. Aerospace engineering, materials engineering, biomedical engineering, ocean engineering, petroleum engineering, mining engineering, nuclear engineering, and manufacturing engineering are examples of these.

The following table shows the number of programs and the number of degrees awarded in 2004/05 in each engineering discipline. Of the 65,183 B.S. degrees awarded, 80 percent were in the top six disciplines, while 20 percent were in the more specialized, non-traditional fields.

ENGINEERING DISCIPLINES RANKED BY NUMBER OF B.S. DEGREES—2004/05 [16]

Discipline	Number of Accredited Programs	B.S. Degrees Awarded in 2004/05	Percent of Total Degrees
Mechanical Engineering	277	14,947	22.9%
Electrical Engineering	284	12,459	19.1
Civil Engineering	233	8,549	13.1
Computer Engineering	183	8,379	12.9
Chemical Engineering	154	4,521	6.9
Industrial Engineering	97	3,482	5.3
Biomedical Engineering	36	2,410	3.7
Aerospace Engineering	61	2,371	3.6
General Engineering	35	1,179	1.9
Materials Sci/Metallurgical Engr	65	840	1.3
Architectural Engineering	14	722	1.1
Agricultural Engineering	43	635	1.0
Systems Engineering	10	570	0.9
Environmental Engineering	46	522	0.8
Marine/Ocean Engineering	16	477	0.7
Engineering Physics/Engr Sci	32	383	0.6

Petroleum Engineering	16	315	0.5
Engineering Management	11	303	0.5
Nuclear Engineering	18	275	0.4
Mining/Mineral/Geological Engr	32	224	0.3
Manufacturing Engineering	24	165	0.3
Ceramic Engineering	6	81	0.1
Other engineering disciplines	12	1,464	2.2
TOTAL	**1,495**	**65,183**	**100.0%**

To find out which of these engineering programs are offered by the 352 universities in the U.S. that have accredited engineering programs, visit the Accreditation Board for Engineering and Technology (ABET) web site at:

http://www.abet.org/accrediteac.asp

There you can search for listings of accredited engineering programs by discipline (e.g., electrical, mechanical, civil, etc.) and by geographical location (region or state).

What follows is an overview of the engineering disciplines. For each of the top six disciplines, more information and details are provided, while for the smaller disciplines briefer descriptions are given.

ELECTRICAL ENGINEERING

Electrical engineering (including computer engineering) is the largest of all engineering disciplines. According to U.S. Department of Labor statistics, of the 1.4 million engineers working with the occupational title of "engineer" in the U.S. in 2005, 353,000 (26 percent) were electrical and computer engineers [9].

Electrical engineers are concerned with electrical devices and systems and with the use of electrical energy. Virtually every industry utilizes electrical engineers, so employment opportunities are extensive. The work of electrical engineers can be seen in the entertainment systems in our homes, in the automobiles we drive, in the computers used by businesses, in numerically-controlled machines used by manufacturing companies, and in the early warning systems used by the federal government to ensure our national security.

Two outstanding sources of information about electrical engineering careers are the Sloan Career Cornerstone Center web site at:

http://www.careercornerstone.org/eleceng/eleceng.htm

and the Institute of Electrical and Electronic Engineers (IEEE) web site at:

http://www.ieee.org/portal/site

As you will see at the IEEE web site, the IEEE is organized into the following 39 technical societies:

Technical Societies of the IEEE

Aerospace and electronic systems	Antennas and propagation
Broadcast technology	Circuits and systems
Communications	Electromagnetic compatibility
Ultrasonics, ferroelectrics, and frequency control	Components, packaging, and manufacturing technology
Computer	Control systems
Consumer electronics	Education
Computational intelligence	Intelligent transportation systems
Dielectrics and electrical insulation	Electron devices
Engineering management	Industrial electronics
Geoscience and remote sensing	Information theory
Industry applications	Lasers and electro-optics
Instrumentation and measurement	Microwave theory and techniques
Magnetics	Oceanic engineering
Nuclear and plasma sciences	Power engineering
Power electronics	Solid state circuits
Professional communication	Reliability
Robotics and automation	Signal processing
Social implications of technology	Systems, man, and cybernetics
Product safety engineering	Vehicular technology
Engineering in medicine and biology	

The listing of IEEE societies should give you an idea of the scope encompassed by the electrical engineering field. Within electrical engineering programs of study, the above 39 technical areas are generally organized under six primary specialties:

Computer Engineering**
Electronics
Communications
Power
Controls
Instrumentation

[**As explained previously, computer engineering will be discussed later in this section as a separate engineering discipline.]

Electronics deals with the design of circuits and electric devices to produce, process, and detect electrical signals. Electronics is rapidly changing and becoming increasingly important because of new advances in microelectronics. Most notable among these has been the doubling of the number of components that can be placed on a given surface area every two years for the past four decades, which has led to much smaller and more powerful devices. Our standard of living has significantly improved due to the advent of semiconductors and integrated circuits (ICs). Semiconductor products include not just digital ICs but also analog chips, mixed-signal (analog and digital integrated) circuits, and radio-frequency (RF) integrated circuits.

Communications involves a broad spectrum of applications from consumer entertainment to military radar. Recent advances in personal communication systems (e.g., cellular telephones, personal assistants, and GPS systems) and video-conferencing, along with technological advances in lasers and fiber optics, are bringing about a revolution in the communications field, opening up possibilities that were not even dreamed of a few years ago: e.g., on-line video-conferencing, international broadcasting of conferences and tutorials, real-time transfer of huge data files, and transmission of integrated voice/data/video files. Wireless communication allows people to communicate anywhere with anyone by voice, e-mail, text, or instant messaging; to send or receive pictures and data; and to access the Internet.

Power involves the generation, transmission, and distribution of electric power. Power engineers are involved with conventional generation systems such as hydroelectric, steam, and nuclear, as well as alternative generation systems such as solar, wind, ocean tides, and fuel cells. Power engineers are employed wherever electrical energy is used to manufacture or produce a product—petrochemicals, pulp, paper, textiles, metals, and rubber, for example. As such, power engineers must have in-depth knowledge about transmission lines, electric motors, and generators.

Controls engineers design systems that control automated operations and processes. Control systems generally compare a measured quantity to a desired standard and make whatever adjustments are needed to bring the measured quantity as close as possible to the desired standard. Control systems are used in regulating the temperature of our buildings, reducing the emissions from our cars and trucks, ensuring the quality of chemical and industrial processes, maintaining reliable electrical output from our power plants, and ensuring highly efficient and fault tolerant voice and data networks. Unmanned aerial vehicles (UAVs) represent a major new and challenging application of controls engineering.

Instrumentation involves the use of electronic devices, particularly transducers, to measure such parameters as pressure, temperature, flow rate, speed, acceleration, voltage, and current. Instrumentation engineers not only conduct such measurements themselves; they also take part in processing, storing, and transmitting the data they collect.

MECHANICAL ENGINEERING

Mechanical engineering is currently the second largest engineering discipline (behind the combined discipline of electrical and computer engineering) in terms of the number of graduates annually and the third largest in terms of the number of employed engineers. According to the U.S. Department of Labor [9], of the 1.4 million engineers in 2005, 221,000 (15.9 percent) were mechanical engineers. Mechanical engineering is also one of the oldest and broadest engineering disciplines.

Mechanical engineers design tools, engines, machines, and other mechanical equipment. They design and develop power-producing machines such as internal combustion engines, steam and gas turbines, and jet and rocket engines. They also design and develop power-using machines such as refrigeration and air-conditioning equipment, robots, machine tools, materials handling systems, and industrial production equipment.

The work of mechanical engineers varies by industry and function. Specialties include, among others, applied mechanics, design, energy systems, pressure vessels and piping, and heating, refrigeration, and air-conditioning systems. Mechanical engineers also design tools needed by other engineers for their work.

The American Society of Mechanical Engineers (ASME) lists 37 technical divisions.

Technical Divisions of the ASME

Applied mechanics	Internal combustion engines
Bioengineering	Fuels and combustion technologies
Fluids engineering	Nuclear engineering
Heat transfer	Power
Tribology	Advanced energy systems
Aerospace	Fluid power systems and technology
Environmental engineering	Noise control and acoustics
Pipeline systems	Computers and information in engineering
Solid waste processing	Information storage & processing systems
Management	Materials handling engineering
Solar energy	Safety engineering and risk analysis
Process industries	Plant engineering and maintenance
Materials	Technology and society
Petroleum	Nondestructive evaluation engineering
Manufacturing engineering	Ocean, offshore, and artic engineering
Design engineering	Pressure vessels and piping
Rail transportation	Electronic and photonic packaging
Textile engineering	Dynamic systems and control
	Microelectromechanical systems (MEMS)

I'm sure this is an overwhelming list, but it is only the "tip of the iceberg." Each of these technical divisions is divided into a number of technical committees. For example, the Applied Mechanics Division is organized into 17 technical committees, each representing subspecialty within the mechanical engineering field. A partial list of these committees will give you some "flavor" of the applied mechanics area:

Composite materials
Computing in applied mechanics
Dynamics and control of structures and systems
Dynamic response of materials
Fracture and failure mechanics

Elasticity
Experimental mechanics
Geomechanics
Materials processing and manufacturing
Fluid mechanics
Instability in solids and structures
Mechanics in biology and medicine
Integrated structures

Within mechanical engineering study, these numerous technical fields and subspecialties are generally grouped into three broad areas:

Energy

Structures and motion in mechanical systems

Manufacturing

Energy involves the production and transfer of energy, as well as the conversion of energy from one form to another. Mechanical engineers in this area design and operate power plants, study the economical combustion of fuels, design processes to convert heat energy into mechanical energy, and create ways to put that mechanical energy to work. Mechanical engineers in energy-related fields also design heating, ventilation, and air-conditioning systems for our homes, offices, commercial buildings, and industrial plants. Some develop equipment and systems for the refrigeration of food and the operation of cold storage facilities; others design "heat exchange" processes and systems to transfer heat from one object to another. Still others specialize in the production of energy from alternative sources such as solar, geothermal, and wind.

The second major area of mechanical engineering study involves the **design of structures** and the **motion of mechanical systems**. Mechanical engineers in these areas contribute to the design of automobiles, trucks, tractors, trains, airplanes, and even interplanetary space vehicles. They design lathes, milling machines, grinders, and drill presses used in the manufacture of goods. They help design the copying machines, faxes, personal computers, and related products that have become staples in our business and home offices. They are involved in the design of the many medical devices, systems, and equipment that help keep us healthy—and, in some cases, alive. Indeed, every piece of machinery that touches our lives, directly or indirectly, has been designed by a mechanical engineer.

Manufacturing, the third area of mechanical engineering study, is the process of converting raw materials into a final product. To take this process from start to finish, a variety of equipment, machinery, and tools is bound to be needed. Designing and building these requisite equipment and machines are what the manufacturing area of mechanical engineering entails. Put simply, mechanical engineers in this area design and manufacture the machines that make machines. They also design manufacturing processes, including automation and robotics, to help make the production of manufactured goods as efficient, cost-effective, and reliable as possible.

If you are interested in learning more about careers in mechanical engineering, check out the ASME student web site:

http://www.asme.org/Communities/Students

CIVIL ENGINEERING

Civil engineering is the currently the third largest engineering field in terms of graduates annually and the second largest in terms of working engineers. According to the U.S. Department of Labor [9], of the 1.4 million engineers working in the U.S. in 2005, 230,000 (16.4 percent) were civil engineers. Civil engineering is the oldest branch of engineering, with major civil engineering projects dating back more than 5,000 years. Today, civil engineers plan, design, and supervise the construction of facilities essential to modern life. Projects range from high-rise buildings to mass transit systems, from airports to water treatment plants, from space telescopes to off-shore drilling platforms.

The American Society of Civil Engineers (ASCE) is organized into 11 technical divisions and councils that develop and disseminate technical information through conferences, book and journals, and policies and standards:

Technical Divisions of ASCE

Aerospace	Cold regions
Computing	Energy
Engineering mechanics	Forensics
Disaster risk management	Lifelines earthquake engineering
Geomatics	Pipelines
Sustainability	

Within civil engineering study, these 11 technical areas are generally organized into seven academic specialties:

Structural engineering

Transportation engineering

Environmental engineering

Water resources engineering

Geotechnical engineering

Surveying

Construction engineering

Structural engineers design all types of structures: bridges, buildings, dams, tunnels, tanks, power plants, transmission line towers, offshore drilling platforms, and space satellites. Their primary responsibility is to analyze the forces that a structure would encounter and develop a design to withstand those forces. A critical part of this design process involves the selection of structural components, systems, and materials that would provide adequate strength, stability, and durability. Structural dynamics is a specialty within structural engineering that accounts for dynamic forces on structures, such as those resulting from earthquakes.

Transportation engineers are concerned with the safe and efficient movement of both people and goods. They thus play key roles in the design of highways and streets, harbors and ports, mass transit systems, airports, and railroads. They are also involved in the design of systems to transport goods such as gas, oil, and other commodities.

Environmental engineers are responsible for controlling, preventing, and eliminating air, water, and land pollution. To these ends, they are typically involved in the design and operation of water distribution systems, waste water treatment facilities, sewage treatment plants, garbage disposal systems, air quality control programs, recycling and reclamation projects, toxic waste cleanup projects, and pesticide control programs.

Water resources is, by its very title, an engineering specialty focused on water-related problems and issues. The work of engineers in this area includes the operation of water availability and delivery systems, the evaluation of potential new water sources, harbor and river development, flood control, irrigation and drainage projects, coastal protection, and the construction and maintenance of hydroelectric power facilities.

Geotechnical engineers analyze the properties of soil and rocks over which structures and facilities are built. From the information their analyses yield, geotechnical engineers are able to predict how the ground material would support or otherwise affect the structural integrity of the planned facility. Their work is thus vital to the design and construction of earth structures (dams and levees), foundations of buildings, offshore platforms, tunnels, and dams. Geotechnical engineers also evaluate the settlement of buildings, stability of slopes and fills, seepage of groundwater, and effects of earthquakes.

Engineers involved in **Surveying** are responsible for "mapping out" construction sites and their surrounding areas before construction can begin. They locate property lines and determine right-of-ways, while also establishing the alignment and proper placement of the buildings to be constructed. Current surveying practice makes use of modern technology, including satellites, aerial and terrestrial photogrammetry, and computer processing of photographic data.

Construction engineers use both technical and management skills to plan and build facilities—such as buildings, bridges, tunnels, and dams—that other engineers and architects designed. They are generally responsible for such projects from start to finish: estimating construction costs, determining equipment and personnel needs, supervising the construction, and, once completed, operating the facility until the client assumes responsibility. Given the breadth of such projects, construction engineers must bc knowledgeable about construction methods and equipment, as well as principles of planning, organizing, financing, managing, and operating construction enterprises.

You can find lots of useful information about civil engineering at the ASCE communities of practice student website at:

http://www.asce.org/community/student/index.cfm

COMPUTER ENGINEERING

Compared to the three previous engineering disciplines we have discussed—electrical engineering, mechanical engineering, and civil engineering—computer engineering a relatively new field. The first accredited computer engineering program in the U.S. was established in 1971 at Case Western Reserve University. Since then, however, computer engineering has experienced rapid growth. It currently ranks fourth in terms of B.S. degrees conferred among engineering disciplines (see table on page 62). And that growth is expected to continue in response to the needs of a world that will become increasingly "computer centered." One indication of the current demand for computer engineers is that the average starting salary for computer engineering graduates in 2005/06 was $53,096 compared to the average for all engineering graduates of $51,465 [13].

Computer engineering, which had its beginnings as a specialty or option within electrical engineering, and continues to rely on much of the same basic knowledge that the EE curriculum teaches, developed into a discipline of its own because of the growing need for specialized training in computer technology. To respond to this need, computer specialists in electrical engineering had to step up their research and course development, which increasingly brought them into contact with computer scientists. Today, although computer engineering and computer science remain separate disciplines, the work of computer engineers and computer scientists is often inseparable—or, more accurately, interdependent. One writer from IEEE aptly explains the relationship between computer engineering and computer science in terms of a "continuum":

> "At one pole is computer science, primarily concerned with theory, design, and implementation of software. It is a true engineering discipline, even though the product is an intangible—a computer program. At the other pole is computer engineering, primarily concerned with firmware (the microcode that controls processors) and hardware (the processors themselves, as well as entire computers). It is not possible, however, to draw a clear line between the two disciplines; many practitioners function to at least some extent as both computer engineers and computer scientists."

While explaining the overlapping nature of the work of computer engineers and scientists, the passage also points out the major difference between them. That is, computer engineers focus more on computer hardware; computer scientists focus more on computer software.

I assume that most of you are already somewhat familiar with these terms. Given their importance in this discussion, however, we'll digress briefly to clarify them. "Hardware" refers to the machine itself: the chips, circuit boards, networks, devices, and other physical components of a computer. "Software" refers to the programs that tell the computer what to do and how to do it. A software program is literally a set of instructions, rules, parameters, and other guidelines, encoded in a special "language" that the hardware can read and then execute. A computer therefore needs both hardware and software, developed in tandem, in order to perform a given function.

As hardware specialists, computer engineers are concerned with the design, construction, assessment, and operation of high-tech devices ranging from tiny microelectronic integrated-circuit chips to powerful systems that utilize those chips and efficient telecommunication systems that interconnect those systems. Applications include consumer electronics (CD and DVD players, televisions, stereos, microwaves, gaming devices) and advanced microprocessors, peripheral equipment (magnetic disks and tapes, optical disks, RAM, ROM, disk arrays, printers and plotters, visual displays, speech and sound software, modems, readers and scanners, keyboards, mouse devices, and speech input systems), systems for portable, desktop and client/server computing, and communications devices (cellular phones, pagers, personal digital assistants). Other applications include distributed computing environments (local and wide area networks, wireless networks, internets, intranets), and embedded computer systems (such as aircraft, spacecraft, and automobile control systems in which computers are embedded to perform various functions). A wide array of complex technological systems, such as power generation and distribution systems and modern processing and manufacturing plans, rely on computer systems developed and designed by computer engineers.

As noted above, however, the work of computer engineers and computer scientists typically involves much crossover. That is, for any given design project, the computer engineer's ability to deliver the appropriate hardware depends on her understanding of the computer scientist's software requirements. As a result, she often participates in the development of the software—and may even create software of her own to support the computer scientist's program. Similarly, the computer scientist's ability to deliver a viable software program depends on his knowledge of hardware. He thus plays a critical role in facilitating the computer engineer's design and development of the necessary physical components, systems, and peripheral devices.

An excellent study "Computer Engineering 2004: Curriculum Guidelines for Undergraduate Degree Programs in Computer Engineering" conduced by a joint task force of the IEEE Computer Society and the Association for Computing Machinery (ACM) lists 18 primary disciplines that make up the body of knowledge for computer engineering [17]. These are:

Algorithms
Computer architecture and organization
Computer systems engineering
Circuits and signals
Database systems
Digital logic
Digital signal processing
Electronics
Embedded systems
Human-computer interaction
Computer networks
Operating systems
Programming fundamentals
Social and professional issues
Software engineering
VLSI design and fabrication
Discrete structures
Probability and statistics

Each of these knowledge areas is further subdivided into sub areas. To give a sense of the scope of the field of computer engineering, for example, the area of "Electronics" includes the following sub areas:

Electronic properties of materials
Diodes and diode circuits
MOS transistors and biasing
MOS logic families
Bipolar transistors and logic families
Design parameters and issues
Storage elements
Interfacing logic families and standard buses
Operational amplifiers
Circuit modeling and simulation
Data conversion circuits
Electronic voltage and current sources
Amplifier design
Integrated circuit building blocks

SYSTEMS ANALYST. Whatever work computer engineers engage in, it is typically generated by some company or government need, which if you recall the engineering design process, leads to a problem definition and specifications. Identifying these needs and initiating design projects to solve them are the responsibility of **systems analysts.** Comprising another fast growing field of computer technology, systems analysts are charged with planning, developing, and selecting new computer systems, or modifying existing programs to meet the needs of an organization. Although their training is in Management Information Systems (as opposed to engineering or computer science), systems analysts are highly computer-literate specialists who work as corporate "watchdogs" to ensure that their company is realizing the maximum benefits from its investment in equipment, personnel, software, and business practices.

COMPUTER SCIENTISTS. Finally, since computer engineers work so frequently—and so closely—with computer scientists, a brief overview of that field would provide a fitting conclusion to our discussion of computer engineering. Computer scientists have already been distinguished as the software experts in the general field of computer technology. As software specialists, their work tends to be highly theoretical, involving extensive, complex applications of math and science principles, algorithms, and other computational processes. However, we have also seen that their theoretical work requires a concomitant knowledge of the many physical components, processes, and functional requirements of computers. The Computing Sciences Accreditation Board (CSAB) begins its definition of the discipline as one that

> "involves the understanding and design of computers and computational processes . . . The discipline ranges from theoretical studies of algorithms to practical problems of implementation in terms of computational hardware and software."

The definition continues, but these two statements alone aptly describe the computer science discipline.

Computer science programs in the U.S. are accredited by the Accreditation Board for Engineering and Technology (ABET) in conjunction with the Computer Science Accreditation Board (CSAB), an organization representing the three largest computer and computer-related technical societies: the Association of Computing Machinery (ACM); the Association for Information Systems (AIS); and the IEEE-Computer Society. During 2004/05, 11,785 computer science degrees were awarded by the 214 universities having accredited computer science programs.

Computer science curricula are comprised of a broad-based core and advanced level subjects. The core provides basic coverage of algorithms, data structures, software design, concepts of programming languages, and computer organization and architecture. Examples of advanced areas include algorithms and data structures, artificial intelligence and robotics, computer networks, computer organization and architecture, database and information retrieval, human-computer interaction, numerical and symbolic computation, operating systems, programming languages, software methodology and engineering, and theory of computation. (It is interesting to note that many of these areas are the same ones listed above for computer engineering. These shared areas only reinforce the overlap and similarities between computer science and computer engineering.)

More information about computing and computer science can be found at the following web sites:

Computer Science Accreditation Board	http://www.csab.org
IEEE Computer Society	http://www.computer.org
Association of Computing Machines	http://www.acm.org
Association for Information Systems	http://plone.aisnet.org

Chemical Engineering

Chemical engineers combine their engineering training with a knowledge of chemistry to transform the laboratory work of chemists into commercial realities. They are most frequently involved in designing and operating chemical production facilities and manufacturing facilities that use chemicals (or chemical processes) in their production of goods.

The scope of the work of chemical engineers is reflected by the industries that employ them—manufacturing, pharmaceuticals, healthcare, design and construction, pulp and paper, petrochemicals, food processing, specialty chemicals, microelectronics, electronic and advanced materials, polymers, business services, biotechnology, and environmental health and safety.

The work of chemical engineers can be seen in a wide variety of products that affect our daily lives, including plastics, building materials, food products, pharmaceuticals, synthetic rubber, synthetic fibers, and petroleum products (e.g., shampoos, soaps, cosmetics, shower curtains, and molded bathtubs).

Chemical engineers also play a major role in keeping our environment clean by creating ways to clean up the problems of the past, prevent pollution in the future, and extend our shrinking natural resources. Many play equally important roles in helping to eliminate world hunger by developing processes to produce fertilizers economically.

The scope of chemical engineering is reflected by the 18 divisions and forums of the American Institute of Chemical Engineers (AIChE):

Divisions and Forums of the AIChE

Catalysis and Reaction Engineering	Environmental
Computational Molecular Science & Engineering	Food, Pharmaceutical & Bioengineering
Computing & Systems Technology	Fuels & Petrochemicals
Transport and Energy Processes	Management
Materials Engineering & Sciences	Nuclear Engineering
Nanoscale Science Engineering	Particle Technology
North American Mixing	Process Development
Safety & Health	Separations
Sustainable Engineering	Forest Products

You can learn more about chemical engineering careers by visiting the AIChE web page at:

http://www.aiche.org/Students/Careers/index.aspx

INDUSTRIAL ENGINEERING

Industrial engineers determine the most effective ways for an organization to use its various resources—people, machines, materials, information, and energy—to develop a process, make a product, or provide a service. Their work does not stop there, however, for they also design and manage the quality control programs that monitor the production process at every step. They also may be involved in facilities and plant design, along with plant management and production engineering.

These multiple responsibilities of an industrial engineer require knowledge not only of engineering fundamentals, but also of computer technology and management practices. At first glance, the industrial engineer might be seen as the engineering equivalent of a systems analyst—except that the industrial engineer plays many more roles and has a much wider window of career opportunities.

Perhaps the single most distinguishing characteristic of industrial engineers is their involvement with the human and organizational aspects of systems design. Indeed, the Institute of Industrial Engineers (IIE) describes industrial engineering as "The People-Oriented Engineering Profession."

Sixty percent of industrial engineers are employed by manufacturing companies, but industrial engineers can be found in every kind of institution (e.g., financial, medical, agricultural, governmental) and commercial field (e.g., wholesale and retail trade, transportation, construction, entertainment, etc.).

Given its breadth of functions in so many areas, industrial engineering has been particularly impacted by recent advances in computer technology, automation of manufacturing systems, developments in artificial intelligence and database systems, management practices (as reflected by the "quality movement"), and the increased emphasis on strategic planning.

The scope of industrial engineering can been seen from the 11 technical societies and divisions of the IIE:

Technical Societies and Divisions of the IIE

Engineering & Management Systems	Quality Control & Reliability Engineering
Work Science	Applied Ergonomics
Computer & Information Systems	Construction
Engineering Economy	Lean
Operations Research	Process Industries
Health Systems	

The unique role of industrial engineering in improving productivity, reducing costs, enhancing customer satisfaction, and achieving superior quality is perhaps reflected by the fact that the IIE includes a *Lean* Division. To learn more about industrial engineering, visit the IIE web site at:

http://www.iienet2.org

OVERVIEW OF OTHER ENGINEERING DISCIPLINES

The following sections provide an overview of each of the more specialized, non-traditional engineering disciplines.

BIOENGINEERING/BIOMEDICAL ENGINEERING. Bioengineering is a wide-ranging field, alternatively referred to as biomedical engineering, which was created some 40 years ago by the merging interests of engineering and the biological/medical sciences. Bioengineers work closely with health professionals in the design of diagnostic and therapeutic devices for clinical use, the design of prosthetic devices, and the development of biologically compatible materials. Pacemakers, blood analyzers, cochlear implants, medical imaging, laser surgery, prosthetic implants, and life support systems are just a few of the many products and processes that have resulted from the team efforts of bioengineers and health professionals.

Biomedical engineering is a field in continual change and creation of new areas due to rapid advancement in technology; however, some of the well established specialty areas within the field are: bioinstrumentation; biomaterials; biomechanics; cellular, tissue and genetic engineering; clinical engineering; medical imaging; orthopedic surgery; rehabilitation engineering; and systems physiology. For more information about careers in biomedical engineering, see the Biomedical Engineering Society (BMES) web page at:

http://www.bmes.org/careers.asp

AEROSPACE ENGINEERING. Aerospace engineers design, develop, test, and help manufacture commercial and military aircraft, missiles, and spacecraft. They also may develop new technologies in commercial aviation, defense systems, and space exploration. In this work, they tend to focus on one type of aerospace product such as commercial transports, helicopters, spacecraft, or rockets. Specialties within aerospace engineering include aerodynamics, propulsion, thermodynamics, structures, celestial mechanics, acoustics, and guidance and control systems.

For more information about aerospace engineering, go to the American Institute of Aeronautics and Astronautics (AIAA) web page at:

http://www.aiaa.org/content.cfm?pageid=214

MATERIALS ENGINEERING/METALLURGICAL ENGINEERING. Materials engineers are generally responsible for improving the strength, corrosion resistance, fatigue resistance, and other characteristics of frequently used materials. The field encompasses the spectrum of materials: metals, ceramics, polymers (plastics), semiconductors, and combinations of materials called composites. Materials engineers are involved in selecting materials with desirable mechanical, electrical, magnetic, chemical, and heat transfer properties that meet special performance requirements. Examples are graphite golf club shafts that are light but stiff, ceramic tiles on the Space Shuttle that protect it from burning up during reentry into the atmosphere, and the alloy turbine blades in a jet engine.

The materials field offers unlimited possibilities for innovation and adaptation through the ability to actually engineer, or create, materials to meet specific needs. This engineering can be carried out at the atomic level through the millions of possible combinations of elements. It can also be done on a larger scale to take advantage of unique composite properties that result from microscopic-scale combinations of metals, ceramics and polymers, such as in fiber reinforcement to make a graphite fishing rod or, on a slightly larger scale, for steel-belted radial tires. Finally, it can be practiced on an even larger scale with bridges, buildings, and appliances.

Metallurgical engineers deal specifically with metals in one of the three main branches of metallurgy—extractive, physical, and mechanical. Extractive metallurgists are concerned with removing metals from ores, and refining and alloying them to obtain useful metal. Physical metallurgists study the nature, structure, and physical properties of metals and their alloys, and design methods for processing them into final products. Mechanical metallurgists develop and improve metal-working processes such as casting, forging, rolling, and drawing.

For more information about materials engineering, go to the Materials Science & Engineering Career Resource Center web site at:

http://www.crc4mse.org/Index.html

ARCHITECTURAL ENGINEERING. Architectural engineers work closely with architects on the design of buildings. Whereas the architect focuses primarily on space utilization and aesthetics, the architectural engineer is concerned with safety, cost, and sound construction methods.

Architectural engineers focus on areas such as the following: the structural integrity of buildings to anticipate earthquakes, vibrations and wind loads; the design and analysis of heating, ventilating and air conditioning systems; efficiency and design of plumbing, fire protection and electrical systems; acoustic and lighting planning; and energy conservation issues.

AGRICULTURAL ENGINEERING. Agricultural engineers are involved in every aspect of food production, processing, marketing, and distribution. Agricultural engineers design and develop agricultural equipment, food processing equipment, and farm structures. Major technical areas of agricultural engineering include food processing, information and electrical technologies, power and machinery, structures, soil and water, forestry, bioengineering, and aqua culture. With their technological knowledge and innovations, agricultural engineers have literally revolutionized the farming industry, enabling farmers today to produce approximately ten times more than what they could just 100 years ago.

For more information on agricultural engineering, visit the American Society of Agricultural and Biological Engineers (ASABE) web site at:

http://www.asabe.org/membership/beengin.html

SYSTEMS ENGINEERING. Systems engineers are involved with the overall design, development, and operation of large, complex systems. Their focus is not so much on the individual components that comprise such systems; rather, they are responsible for the integration of each component into a complete, functioning "whole." Predicting and overseeing the behavior of large-scale systems often involves knowledge of advanced mathematical and computer-based techniques, such as linear programming, queuing theory, and simulation.

For more information on systems engineering, visit the International Council on Systems Engineering (INCOSE) web site at:

http://www.incose.org

Click on "Education & Careers." Then click on "Careers in SE."

ENVIRONMENTAL ENGINEERING. Environmental engineers, relying heavily on the principles of biology and chemistry, develop solutions to environmental problems. Environmental engineers work in all aspects of environmental protection including air pollution control, industrial hygiene, radiation protection, hazardous waste management, toxic materials control, water supply, wastewater management, storm water management, solid waste disposal, public health, and land management.

Environmental engineers conduct hazardous-waste management studies in which they evaluate the significance of the hazard, offer analysis on treatment and containment, and develop regulations to prevent mishaps.

They design municipal water supply and industrial wastewater treatment systems. They conduct research on proposed environmental projects, analyze scientific data, and perform quality control checks. They provide legal and financial consulting on matters related to the environment. For information about environmental engineering, see the American Academy of Environmental Engineers (AAEE) web site at:

http://www.aaee.net/Website/Careers.htm

MARINE ENGINEERING/OCEAN ENGINEERING/NAVAL ARCHITECTURE. Naval Architects, Marine Engineers, and Ocean Engineers design, build, operate, and maintain ships and other waterborne vehicles and ocean structures as diverse as aircraft carriers, submarines, sailboats, tankers, tugboats, yachts, underwater robots, and oil rigs. These interrelated professions address our use of the seas and involve a variety of engineering and physical science skills, spanning disciplines that include hydrodynamics, material science, and mechanical, civil, electrical, and ocean engineering

Marine Engineers are responsible for selecting ships' machinery, which may include diesel engines, steam turbines, gas turbines, or nuclear reactors, and for the design of mechanical, electrical, fluid, and control systems throughout the vessel. Some marine engineers serve aboard ships to operate and maintain these systems. Ocean Engineers study the ocean environment to determine its effects on ships and other marine vehicles and structures. Ocean engineers may design and operate stationary ocean platforms, or manned or remote-operated sub-surface vehicles used for deep sea exploration. Naval Architects are involved with basic ship design, starting with hull forms and overall arrangements, power requirements, structure, and stability. Some naval architects work in shipyards, supervising ship construction, conversion, and maintenance.

For more information on marine engineering, ocean engineering, and naval architecture, visit the Society of Naval Architects & Marine Engineers web site at:

http://www.sname.org/careers.htm

PETROLEUM ENGINEERING. Petroleum engineers work in all capacities related to petroleum (gas and oil) and its byproducts. These include designing processes, equipment, and systems for locating new sources of oil and gas; sustaining the flow of extant sources; removing, transporting, and storing oil and gas; and refining them into useful products.

For more information about petroleum engineering, go to the Society of Petroleum Engineering (SPE) web site at:

http://www.spe.org

Click on "Career Development" and then click on "About Petroleum Engineers."

NUCLEAR ENGINEERING. Nuclear engineers are concerned with the safe release, control, utilization, and environmental impact of energy from nuclear fission and fusion sources. Nuclear engineers are involved in the design, construction, and operation of nuclear power plants for power generation, propulsion of nuclear submarines, and space power systems. Nuclear engineers are also involved in processes for handling nuclear fuels, safely disposing radioactive wastes, and using radioactive isotopes for medical purposes.

MINING ENGINEERING/GEOLOGICAL ENGINEERING. The work of mining and geological engineers is similar to that of petroleum engineers. The main difference is the target of their efforts. That is, mining and geological engineers are involved in all aspects of discovering, removing, and processing minerals from the earth. The mining engineer designs the mine layout, supervises its construction, and devises systems to transport minerals to processing plants. The mining engineer also devises plans to return the area to its natural state after extracting the minerals.

MANUFACTURING ENGINEERING. Manufacturing engineers are involved in all aspects of manufacturing a product. These include studying the behavior and properties of required materials, designing appropriate systems and equipment, and managing the overall manufacturing process.

To learn more about manufacturing engineering, visit the Society of Manufacturing Engineers (SME) web site at:

http://www.manufacturingiscool.com

CERAMIC ENGINEERING. Ceramic engineers direct processes that convert nonmetallic minerals, clay, or silicates into ceramic products. Ceramic engineers work on products as diverse as glassware, semiconductors, automobile and aircraft engine components, fiber-optic phone lines, tiles on space shuttles, solar panels, and electric power line insulators.

REFLECTION

Reflect on the 20 engineering disciplines described in this section. Have you already decided which one you will major in? Why did you choose it? If you haven't yet chosen a specific engineering discipline, which one is the most appealing to you at this point in time? What specifically about it do you find appealing?

2.7 ENGINEERING JOB FUNCTIONS

Another way to understand the engineering profession is to examine engineers from the perspective of the work they do or the job functions they perform. For example, an electrical engineer could also be referred to as a *design* engineer, a *test* engineer, or a *development* engineer—depending on the nature of his or her work. Following is a description of the nine main engineering job functions.

ANALYSIS

The **analytical engineer** is primarily involved in the mathematical modeling of physical problems. Using the principles of mathematics, physics, and engineering science—and making extensive use of engineering applications software—the analytical engineer plays a critical role in the initial stage of a design project, providing information and answers to questions that are easy and inexpensive to obtain. Once the project moves from the conceptual, theoretical stage to the actual fabrication and implementation stage, changes tend to be time-consuming and costly.

DESIGN

The **design engineer** converts concepts and information into detailed plans and specifications that dictate the development and manufacture of a product. Recognizing that many designs are possible, the design engineer must consider such factors as production cost, availability of materials, ease of production, and performance requirements. Creativity and innovation, along with an analytic mind and attention to detail, are key qualifications for a design engineer.

TEST

The **test engineer** is responsible for developing and conducting tests to verify that a selected design or new product meets all specifications. Depending on the product, tests may be required for such factors as structural integrity, performance, or reliability—all of which must be performed under all expected environmental conditions. Test engineers also conduct quality control checks on existing products.

DEVELOPMENT

The **development engineer,** as the title indicates, is involved in the development of products, systems, or processes. The context in which such "development" occurs, however, can vary considerably. Working on a specific design project, the development engineer acts as a kind of "intermediary" between the design and test engineers. He helps the design engineer to formulate as many designs as possible that meet all specifications and accommodate any constraints. Once a design is selected, the development engineer oversees its fabrication—usually in the form of a prototype or model. The results of his collaboration with the design engineer and subsequent supervision of the prototype's fabrication are bound to affect the kind and amount of testing the test engineer will then need to conduct.

In a more general context, the development engineer is instrumental in turning concepts into actual products or applying new knowledge to improve existing products. In this capacity, he is the "D" in "R&D," which, as you probably know, stands for the Research and Development arm of many companies. Here, the development engineer is responsible for determining how to actualize or apply what the researcher discovers in the laboratory, typically by designing, fabricating, and testing prototypes or experimental models.

SALES

The **sales engineer** is the liaison person between the company and the customer. In this role, the sales engineer must be technically proficient in order to understand the product itself and the customer's needs. That means she must be able to explain the product in detail: how it operates, what functions it can perform, and why it will satisfy the customer's requirements. She also needs to maintain a professional working relationship as long as the customer is using her company's products. She must be able to field questions about the product, explain its features to new users, and arrange prompt, quality service should the customer experience problems with the product. Obviously, along with solid technical knowledge, the sales engineer must possess strong communication skills and related "people" skills.

RESEARCH

The work of the **research engineer** is not unlike that of a research scientist in that both are involved in the search for new knowledge. Where they differ is the purposes that motivate their work. Scientific researchers are generally interested in the new knowledge itself: what it teaches or uncovers about natural phenomena. Engineering researchers are interested in ways to apply the knowledge to engineering practices and principles. Research engineers thus explore mathematics, physics, chemistry and engineering sciences in search of answers or insights that will contribute to the advancement of engineering.

Given the nature and demands of their work, research engineers usually need to have an advanced degree in their field. Indeed, most positions available in engineering research require a Ph.D.

MANAGEMENT

If you are successful as an engineer and have strong leadership skills, within a few years of graduation you could very well move into management. Opportunities exist primarily in two areas: **line management** and **project management**.

In a company, the technical staff is generally grouped into an engineering "line organization." At the base of this "line" are units of ten to 15 engineers, who are managed by a unit supervisor. At the next level up the line, these units report to a group manager. This organizational line continues up to department managers, a chief engineer or engineering vice president, and finally the president. Often the president of a technical company is an engineer who worked his or her way up through the line organization.

Project management is a little different, as the personnel are organized according to a specific project or assignment. At the head of each project is a project manager. For a small project, one manager is usually sufficient to oversee the entire project; for a larger project, the project manager is assisted by a professional staff, which can range from one to several hundred people. The overall responsibility of the project manager and staff is to see that the project is completed successfully, on time, and within budget.

CONSULTING

The work of the **consulting engineer** differs from that of all other engineers in that a consulting engineer performs services for a client on a contractual basis. Some consulting engineers are self-employed, while others work for consulting firms that "hire out" their engineers to companies that either lack the expertise the consulting engineer can provide or want an outside evaluation of their organization's performance. Depending on the client's specific needs, the consulting engineer's work can vary considerably. Investigations and analyses; preplanning, design and design implementation; research and development; construction management; and recommendations regarding engineering-related problems are just a few examples.

The time a consulting engineer puts into each assignment also can vary. Sometimes the work can be done in a day; other times it can require weeks, months, or even years to complete. Last, engineering consulting is increasingly becoming a global enterprise. Both the public and private sectors of developing nations have growing technological needs and so turn to U.S. consulting firms for technical assistance. If the diversity of work and opportunity to travel catch your interest, a career in engineering consulting could be for you.

TEACHING

The **engineering professor** has three primary areas of responsibility: teaching, research, and service. Teaching includes not only classroom instruction, but also course and curriculum development, laboratory development, and the supervision of student projects or theses. Research involves the pursuit of new knowledge, which is then disseminated throughout the professional engineering community by papers published in engineering journals, presentations at scholarly meetings, textbooks, and software. The research demands of the engineering educator also include success in generating funds to support research projects, as well as participation in professional societies. "Service" is a catch-all term that refers to the many other functions expected of engineering professors. These include such activities as community involvement, participation in faculty governance, public service, and consulting.

The Ph.D. degree in engineering is virtually mandatory to qualify for a full-time position on an engineering faculty at a four-year institution, while an M.S. degree is generally sufficient for a teaching position at a community college. More information about academic careers in engineering can be found in Reference 18.

REFLECTION

Reflect on the nine engineering job functions described in this section. Which of them is the most appealing to you? Analysis? Design? Test? Development? Sales? Research? Could you see yourself in management? Could you see yourself as an engineering professor?

2.8 EMPLOYMENT OPPORTUNITIES

When you graduate in engineering, you will face a number of choices. The first will be whether you want to continue your education full time or seek work as a practicing engineer. If you elect to continue your education, you next need to decide whether you want to pursue your M.S. degree in engineering or do graduate work in another field, such as business administration, law, or medicine. Opportunities for graduate study will be discussed in Chapter 8.

If you decide to seek a full-time engineering position, many opportunities and choices will await you. The field of engineering practice is so vast and the job opportunities so varied, you may well need to devote a substantial amount of time to understand fully the opportunities and areas of practice available to you.

Rather than waiting until you graduate to learn about the many opportunities that engineering has to offer, you should make this an objective early on in your engineering studies. Besides saving time and energy when you launch your job search later, knowing NOW about the many areas in which engineers are needed and the diverse opportunities that await you will be a strong incentive for you to complete your engineering studies.

Let's start, then, with a "big picture" view of the major areas in which most engineers work. The table below, which lists these areas, along with numbers and percentages tabulated for 2003, provides this view [19].

Employed Individuals with Engineering Degrees – 2003

Employment Area	Number	Percentage
Business/Industry	1,848,900	79.2%
Federal Government	136,600	5.9%
State/Local Government	124,600	5.3%

Educational Institutions	126,500	5.4%
Self-Employed	96,500	4.1%
Total	**2,333,200**	**100%**

As you can see, the first area, "Business/Industry," is clearly the largest, employing 79.2 percent of engineers. You should know, however, that "industry" is a blanket term for two distinct categories: (1) manufacturing; and (2) non-manufacturing (service). *Manufacturing* is involved in converting raw materials into products, while *non-manufacturing* concerns the delivery of services. Government, the next highest area, employing 11.2 percent of engineers, has needs for engineers at the local, state, and federal levels. Following business, industry, and government comes educational institutions, which employ a large number of engineers, both as engineering professors and as researchers in university-operated research laboratories. Finally, there is a small but significant area of self-employed engineers, most of whom are consulting engineers.

You will also note that the total number of employed individuals having engineering degrees is 2.33 million, whereas we previously stated that the number of people working with the title of "engineer" was 1.4 million (see page 47). The difference in these numbers reflects the fact that about one-third of individuals having engineering degrees work in positions that do not have the title "engineer."

ORGANIZATION OF INDUSTRY IN THE UNITED STATES

If almost 80 percent of engineers work in the "Business/Industry" area, it is likely that you, too, will find yourself working in this area. Although we briefly mentioned the two categories into which industry is divided (manufacturing and non-manufacturing), we have barely scratched the surface of this huge, complex field. For a more detailed, comprehensive perspective on the many, many diverse fields that comprise U.S. business and industry, **The North American Industry Classification System (NAICS)** [20] is the best resource available.

Developed and maintained by the U.S. government, the NAICS system dissects the monolithic term "business and industry" into 1,179 national industries, each identified by a six-digit classification code. It then lists all the products or services that each national industry provides.

To give you an idea of how NAICS works, I randomly selected the following ten of the 1,179 national industries in the NAICS classification system:

211111	Crude petroleum and natural gas extraction
221112	Fossil-fuel electric power generation
237310	Highway, street, and bridge construction
325611	Soap and other detergent manufacturing
334510	Electromedical and electrotherapeutic apparatus manufacturing
335311	Power, distribution, and specialty transformer manufacturing
335921	Fiber-optic cable manufacturing
336414	Guided missile and space vehicle manufacturing
517212	Cellular and other wireless telecommunications
541330	Engineering services

The first two digits designate a major "Economic Sector," and the third digit designates an "Economic Subsector." For example, all of the industry subgroups above starting with 33 are part of the "Manufacturing" Economic Sector. The two national industries in the list whose first three digits are 335 fall under the "Electrical Equipment, Appliance, and Component Manufacturing" Economic Subsector. The remaining digits of each six-digit classification code further subdivide the subsectors into: industry groups (4 digits); NAICS industries (5 digits); and national industries (6 digits).

As an example, consider the "Economic Sector,"

33 – Manufacturing

Under this economic sector, there are eight "Economic Subsectors,"

331 – Primary metal manufacturing
332 – Fabricated product metal manufacturing
333 – Machinery manufacturing
334 – Computer and electronic product manufacturing
335 – Electrical equipment, appliance, and component manufacturing
336 – Transportation equipment manufacturing
337 – Furniture and related product manufacturing
339 – Miscellaneous manufacturing

Take, one of the "Economic Subsectors," for example, NAICS 334, Computer and Electronic Product Manufacturing. Under this economic subsector, there are six "Industry Groups."

3341 – Computer and peripheral equipment manufacturing
3342 – Communications equipment manufacturing
3343 – Audio and video equipment manufacturing
3344 – Semiconductor and other electronic component manufacturing
3345 – Navigational, measuring, electromedical, and control instrument manufacturing
3346 – Manufacturing and reproducing magnetic and optical media

Take one of the "Industry Groups," for example NAICS 3345, Navigational, measuring, electromedical, and control instrument manufacturing. Under this industry group there are ten national industries:

334510 – Electromedical/Electrotherapeutic Apparatus Manufacturing
334511 – Search, Detection, Navigation, Guidance, Aeronautical, and Nautical System and Instrument Manufacturing
334512 – Automatic Environmental Control Manufacturing for Residential, Commercial, and Appliance Use
334513 – Instruments and Related Products Manufacturing for Measuring, Displaying, and Controlling Industrial Process Variables
334514 – Totalizing Fluid Meter and Counting Device Manufacturing
334515 – Instrument Manufacturing for Measuring and Testing Electricity and Electrical Signals
334516 – Analytical Laboratory Instrument Manufacturing
334517 – Irradiation Apparatus Manufacturing
334518 – Watch, Clock, and Part Manufacturing
334519 – Other Measuring and Controlling Device Manufacturing

And if you pick just one of these ten national industries, say for example 334510 – Electromedical/Electrotherapeutic Apparatus Manufacturing, you'll find listed more than 50 major products such as magnetic resonance imaging equipment, medical ultrasound equipment, pacemakers, hearing aids, electrocardiographs, and electromedical endoscopic equipment.

I hope this has given you an idea of the enormity of U.S. business and industry and the tools you need to access that industry. You can explore the North American Industry Classification System on your own on line at: http://www.census.gov. Click on "NAICS," followed by "NAICS Search." Then enter one of two types of keywords:

(1) a product or service (e.g., "fiber optic")

(2) an NAICS classification (e.g., "NAICS 541330")

EXERCISE

Go to the U.S. Bureau of Census web page at: http://www.census.gov. Click on "NAICS" and then conduct a "NAICS search" on the manufacturing economic sector "NAICS 33." Scroll down until you find a national industry you would be interested in working in. Click on the six-digit code for that national industry to see a listing of the products manufactured by that industry. Pick one of the products and do an Internet search using a search engine such as Google to identify the companies that compete in the marketplace for that product. Pick one of the companies and go to that company's web site and see if you can identify job listings for engineers.

Because, as we learned earlier, almost 80 percent of engineers work in business and industry, it would be good to know how many of these engineers work in various areas of both "manufacturing" and "service" industries

The economic sectors or economic subsectors that employ the largest number of engineers are shown in the following table.

Economic Sector or Subsector	NAICS Code	Number of Engineers	% of All Engineers
Manufacturing Subsector			
Computer and electronic product	334	189,610	13.1%
Transportation equipment	336	119,130	8.4
Machinery	333	62,050	4.4
Fabricated metal product	332	31,810	2.3
Chemical	325	27,530	1.9
Electrical equipment, appliance, and component	335	22,960	1.6

Service Sector			
Professional, scientific, and technical services	54	402,070	28.5%
Information	51	46,190	3.3
Construction	23	37,610	2.7
Wholesale Trade	42	37,390	2.7
Administrative and support	56	36,710	2.6
Management of companies and enterprises	55	35,300	2.5
Utilities	22	27,720	2.0
Mining	21	18,260	1.3

The following sections briefly describe these economic sectors and subsectors.

Manufacturing Subsectors Employing Highest Numbers of Engineers

Computer and Electronic Product Manufacturing. These industries are engaged in the manufacture of computers, computer peripherals, communication equipment, and related electronic equipment. Their manufacturing processes differ fundamentally from those of other machinery and equipment in that the design and use of integrated circuits and the application of highly specialized miniaturization technologies are common elements in the manufacturing processes of computer and electronic products.

Transportation Equipment Manufacturing. These industries produce equipment and machinery needed for transporting people and goods. Their manufacturing processes are similar to those used in most other machinery manufacturing establishments—bending, forming, welding, machining, and assembling metal or plastic parts into components and finished products. Evidence of the equipment and machinery manufactured in this subsector can be found in such transportation products as motor vehicles, aircraft, guided missiles and space vehicles, ships, boats, railroad equipment, motorcycles, bicycles, and snowmobiles.

MACHINERY MANUFACTURING. These industries design and produce products that require mechanical force to perform work. Both general-purpose machinery and machinery designed to be used in a particular industry are included in this subsector. Examples of general-purpose machinery include heating, ventilation, air-conditioning, and commercial refrigeration equipment; metalworking machinery; and engine, turbine, and power transmission equipment. Three categories of special purpose machinery are included: agricultural, construction, and mining machinery; industrial machinery; and commercial and service industry machinery.

FABRICATED METAL PRODUCT MANUFACTURING. These industries transform metal into intermediate or end products using forging, stamping, bending, forming, and machining to shape individual pieces of metal. They also use processes, such as welding and assembling, to join separate parts together. Examples of products include hand tools, kitchen utensils, metal containers, springs, wire, plumbing fixtures, firearms, and ammunition.

CHEMICAL MANUFACTURING. These industries manufacture three general classes of products: (1) basic chemicals, such as acids, alkalies, salts, and organic chemicals; (2) chemical products to be used in further manufacture, such as synthetic fibers, plastics materials, dry colors, and pigments; and (3) finished chemical products to be used for human consumption, such as drugs, cosmetics, and soaps; or products to be used as materials or supplies in other industries, such as paints, fertilizers, and explosives.

ELECTRICAL EQUIPMENT, APPLIANCE, AND COMPONENT MANUFACTURING. These industries manufacture products that generate, distribute, and use electrical power. Electric lighting equipment, household appliances, electric motors and generators, batteries, and insulated wire and wiring devices are but a few of the many products that come under this manufacturing subsector.

SERVICE SECTORS EMPLOYING HIGHEST NUMBERS OF ENGINEERS

PROFESSIONAL, SCIENTIFIC, AND TECHNICAL SERVICES. This sector includes industries from three large areas, only one of which—"Technical Services"—applies to engineering. Under "Technical Services," however, NAICS includes a broad, varied list of both engineering and computer services. Engineering services may involve any of the following: provision of advice (i.e., engineering consulting), preparation of feasibility studies, preparation of preliminary plans and designs, provision of technical services during the construction or implementation stages of a project, inspection and evaluation of completed projects, and related services. Computer services are equally varied, including activities such as programming, computer-integrated systems design, data preparation and processing, information retrieval, facilities management, as well as computer leasing, maintenance, and repair.

INFORMATION. These industries are engaged in three main processes: (1) producing and distributing information and cultural products; (2) providing the means to transmit or distribute these products, along with data or communications; and (3) processing data. Subsectors under this sector include publishing industries, motion picture and sound recording industries, broadcasting and telecommunications, and information and data processing services.

CONSTRUCTION. These industries cover three broad areas of construction: (1) building construction, such as dwellings, office buildings, commercial buildings, stores, and farm buildings; (2) heavy construction other than buildings, such as highways, streets, bridges, sewers, railroads, irrigation projects, flood control projects, and marine construction; and (3) special trades for heavy construction such as painting, electrical work, plumbing, heating, air-conditioning, roofing, and sheet metal work.

WHOLESALE TRADE. Wholesale trade includes: (1) merchant wholesalers who take title to the goods they sell; (2) sales branches or offices maintained by manufacturing, refining, or mining enterprises; and (3) agents, merchandise or commodity brokers, and commission merchants. The merchandise includes the outputs of agriculture, mining, manufacturing, and certain information industries, such as publishing.

ADMINISTRATION AND SUPPORT. These industries perform routine support activities for the day-to-day operations of other organizations. These essential activities are often undertaken in-house by establishments in many sectors of the economy. Activities performed include: office administration, hiring and placing of personnel, document preparation and similar clerical services, solicitation, collection, security and surveillance services, cleaning, and waste disposal services.

MANAGEMENT OF COMPANIES AND ENTERPRISES. This sector comprises (1) establishments that hold the securities of (or other equity interests in) companies and enterprises for the purpose of owning a controlling interest or influencing management decisions or (2) establishments (except government establishments) that administer, oversee, and manage establishments of the company or enterprise and that normally undertake the strategic or organizational planning and decision making role of the company or enterprise. Presumably, engineers who are provided to companies on a contract basis are included in this economic sector.

UTILITIES. These industries are engaged in providing electric power, natural gas, steam, water, and sewage removal. Providing electric power includes generation, transmission, and distribution, while natural gas only involves distribution. Supplying steam includes provision and/or distribution; supplying water involves treatment and distribution. Sewage removal includes collection, treatment, and disposal of waste through sewer systems and sewage treatment facilities.

MINING. These industries extract naturally occurring mineral solids, such as coal and ores; liquid minerals, such as crude petroleum; and gases, such as natural gas. The term "mining" is used in the broad sense to include quarrying, well operations, beneficiating (e.g., crushing, screening, washing, and flotation), and other preparatory functions customarily done at the mine site.

2.9 IMPORTANT FIELDS FOR THE FUTURE

We are in a period of intense change. One way to underscore this is to reflect on the fact that none of the following existed as recently as 1980 (25 years ago).

1. The Internet
2. Cell phone
3. Personal computers
4. Fiber optics
5. E-mail
6. Commercialized Global Positioning System (GPS)
7. Portable computers
8. Memory storage discs
9. Consumer level digital camera
10. Radio frequency ID tags
11. Micro-Electrical-Mechanical Systems (MEMS)
12. DNA fingerprinting
13. Air bags
14. ATM
15. Advanced batteries

16. Hybrid car
17. Organic Light Emitting Diodes (OLEDs)
18. Display panels
19. HDTV
20. Space shuttle
21. Nanotechnology
22. Flash memory
23. Voice mail
24. Modern hearing aids
25. Short Range, High Frequency Radio

This list of the top 25 non-medically related technology innovations of the past 25 years was developed by a panel of technology leaders assembled by the Lemelson-MIT Program at MIT in 2005. These advances have not only created exciting new opportunities for engineers, they have resulted in a "flattening" of the world.

I would encourage you to read Thomas L. Friedman's excellent book *The World is Flat: A Brief History of the Twenty-First Century* [21]. Doing so will help you understand the world you will be living and working in. In the book, Friedman explores the political and technological changes that have flattened the world and made it a smaller place. From the fall of the Berlin Wall to the explosion of the Internet to the *dot com* bubble and bust and outsourcing of jobs to India and China, globalization has evened the playing field for many emerging economies.

The following is a list of some of the major events and changes that will influence your future as an engineer:

Major Events and Changes Affecting the Future

The fall of the Berlin Wall
Advances in computer technology
Advances in communications
The knowledge and information explosion
Globalization (outsourcing, offshoring)
Increased focus on the environment
Events of September 11, 2001
World population explosion

Understanding these changes can help you prepare for the engineering fields that will be particularly important and "in demand" in the years ahead. The following are ten areas of technology that the National Science Foundation (NSF) [22,23] and the National Academy of Engineering (NAE) [24] have targeted for rapid development in the future.

MANUFACTURING FRONTIERS

Manufacturing is the foundation of the U.S. economy. There is now an unprecedented opportunity to accelerate the application of new knowledge and advanced technologies to dramatically improve the manufacturing capabilities of U.S. industries. Important technologies include "intelligent" manufacturing, advanced fabrication and processing methods, integrated computer-based tools for product design and manufacturing, systems and processes to prevent pollution and to minimize resource waste, and "total quality management" systems and processes.

INFORMATION AND COMMUNICATION SYSTEMS

Advances in **information and communication technologies** are key to U.S. economic growth and competitiveness—as well as to our national defense. The transition from analog to digital processing is enabling the U.S. to regain its competitiveness in consumer electronics. We can hardly imagine a world without computers, cellular phones, e-mail, I-pods, and the Internet. Information and communication technologies will continue to bring about major changes in health care delivery systems, advanced manufacturing technology, civil infrastructure systems, and approaches to learning in engineering education.

SMART AND ENGINEERED MATERIALS

Improvement in the manufacture and performance of **materials** will enhance our quality of life, national security, industrial productivity, and economic growth. New materials will be created that feature precisely tailored properties and enhanced performance. Examples of these materials are biodegradable polymers, high-temperature ceramics, and durable materials for artificial limbs and joints. Work is underway on developing "smart" materials that are designed to react to changes in their environment, such as materials that can sense motion and counter it, thus damping vibration, or change shape or viscosity in response to stress, temperature or electrical activity, but "remember" their original configurations. In the future, smart materials will be employed in creating the coming generation of compact, low-power sensors that can detect toxic chemicals, bio-hazards or radiation, as well as dozens of other stimuli.

BIOENGINEERING

Bioengineering is expected to have a profound impact on health care, agriculture, energy, and environmental management. Major areas of activity focus on developing better ways to manufacture and mass-produce pharmaceuticals, better safeguards for the environment (including the eradication of past problems), and better means of improving, restoring, and preserving human health. Exciting and developing fields within bioengineering include: tissue engineering, biomechanics, rehabilitation engineering, bioinformatics, neural engineering, and biomedical imaging.

CRITICAL INFRASTRUCTURE SYSTEMS

Our **critical infrastructure system** is the framework of networks and systems that provides a continual flow of goods and services essential to the defense and economic security of the United States. These include: agriculture and food; water; public health; emergency services; the defense industrial base; information and telecommunications; energy and power; transportation; and banking and finance. The facilities, systems, and functions that comprise our critical infrastructures are highly sophisticated, interdependent, and complex. Engineering will play a major role in maintaining and improving this infrastructure. Of particular concern is the civil engineering infrastructure—consisting of roads, bridges, rail networks, airports, sewage treatment plants, deep-water ports, municipal water systems, and energy generation and transmission systems—which is both deteriorating and inadequate to meet growing demands. Rebuilding and expanding this infrastructure will involve new designs, more durable materials, network systems with better controls and communications, and improved management processes.

HOMELAND SECURITY

Since the events of September 11th, the prospect of continued terrorism in the U.S. and abroad and the issues related to **homeland security** have been of paramount importance to policy makers and the public. Engineers will play an important role in the process of improving our security. Technologies originally developed for other purposes are now being explored for counter-terrorism, as in the application of optical spectroscopy to the detection chemical weapons. New technologies, such as face recognition software, are also being created. Moreover, new technologies developed outside of the arena of terrorism concerns, such as wireless computing, present a new set of security issues.

IMPROVED HEALTH CARE DELIVERY

There is a critical national need to contain the costs of health care, while also **improving the quality of and access to health care**. Engineering will help meet this need by developing new ways to increase productivity in hospitals, new technologies for the delivery of care outside of hospitals, improved materials for use in implants or external devices to increase longevity, and improved information and communication systems to expedite access to health care and increase patients' independence.

NANOTECHNOLOGY

Nanotechnology is a catch-all phrase for materials and devices that operate at the nanoscale. In the metric system of measurement, "Nano" equals a billionth and therefore a nanometer is one-billionth of a meter. Two main approaches are used in nanotechnology: one is a "bottom-up" approach where materials and devices are built up atom by atom, the other a "top-down" approach where they are synthesized or constructed by removing existing material from larger entities. Increasingly control of matter and energy at the molecular level is already leading to revolutionary breakthroughs in such critical fields as advanced computing, communications, materials development, and medicine. In communications, nanostructures are dramatically reducing the size of signal processing components and have led to new abilities to control light beams. In medicine, ultra-miniaturized sensors and fluid channels are ushering in a new era of tiny diagnostic and detection devices. Nanofabrication of miniature electronic components may revolutionize information processing.

ADVANCED ENVIRONMENTAL TECHNOLOGY

Our nation's industrial development and economic growth will require solutions to extremely **complex environmental problems**, such as finding new ways to manage natural resources, while stepping up the production of goods and services. Engineers will play important roles in creating technologies and processes that will remediate existing problems and prevent future problems. One example of such technologies is environmentally sound extraction/production systems that minimize or prevent waste and contamination. Engineers also will be involved in promoting a better understanding of the relationships between human needs and the environment.

Sensors and Control Systems

Because of technological advances in the areas of signal processing, communications, bioengineering, computing power, and sensor technology, the field of **sensors and control systems** can provide new, more accurate, less expensive, and more efficient control solutions to existing and novel problems. We are moving toward autonomous underwater, land, air, and space vehicles; highly automated manufacturing; intelligent robots; highly efficient and fault tolerant voice and data networks; reliable electric power generation and distribution; seismically tolerant structures; and highly efficient fuel control for a cleaner environment.

2.10 Engineering as a Profession

When you receive your B.S. degree in engineering, you will join the engineering profession. Engineering may be considered a profession insofar as it meets the following characteristics of a learned professional group [25]:

- Knowledge and skill in specialized fields above that of the general public
- A desire for public service and a willingness to share discoveries for the benefit of others
- Exercise of discretion and judgment
- Establishment of a relation of confidence between the professional and client or employer
- Self-imposed (i.e., by the profession) standards for qualifications
- Acceptance of overall and specific codes of conduct
- Formation of professional groups and participation in advancing professional ideals and knowledge
- Recognition by law as an identifiable body of knowledge

As an engineering professional, you will have certain rights and privileges, as well as certain responsibilities and obligations. As described above, you will be responsible for serving the public, sharing your discoveries for the benefit of others, exercising discretion and judgment, maintaining confidentiality with clients and employers, and accepting specific codes of conduct.

As an engineering professional, you will have the legal right to represent yourself using the title of engineer. You will be eligible to participate in professional organizations. And you will have the right to seek registration as a *Professional Engineer.*

PROFESSIONAL REGISTRATION

You can formalize your status as a professional by seeking registration as a *Professional Engineer* (P.E.). Professional registration is an impressive credential, and you will find the title ***P.E.*** proudly displayed on the business cards of engineers who have acquired that status. For most engineers, professional registration is optional. However, in certain fields of work that involve public safety, professional registration may be mandatory. Approximately 30 percent of all practicing engineers are registered. The percentage is much higher for civil engineers because of the nature of their work.

Professional registration is handled by the individual states, each of which has a registration board. Although the requirements and procedures differ somewhat from state to state, they are generally fairly uniform due to the efforts of the National Council of Examiners for Engineers and Surveyors (NCEES). For details about the process of becoming a registered Professional Engineer, visit the NCEES web page:

http://www.ncees.org

State boards are responsible for evaluating the education and experience of applicants for registration, administering an examination to those applicants who meet the minimum requirements, and granting registration to those who pass the examination.

Although registration laws vary, most boards require four steps:

1. Graduation from a four-year engineering program accredited by the Accreditation Board for Engineering and Technology (ABET)
2. Passing the Fundamentals of Engineering (FE) examination
3. Completing four years of acceptable engineering practice
4. Passing the Principles and Practice of Engineering (PE) examination

Once you complete these four steps, you will become licensed as a Professional Engineer in the state in which you wish to practice, and will be certified to use the prestigious "P.E." designation after your name. Most states provide for reciprocal licensure, so that once you become licensed in one state, you can become licensed in other states without further examination.

The Fundamentals of Engineering Exam. The Fundamentals of Engineering Exam (FE) is administered each year in April and October. The FE exam is an eight-hour multiple-choice exam. The four-hour morning session is common to all engineering disciplines and is comprised of 120 one-point questions covering the following topics:

Mathematics (15%)
Engineering probability and statistics (7%)
Chemistry (9%)
Engineering economics (8%)
Computers (7%)
Ethics and business practice (7%)
Engineering mechanics (statics and dynamics) (10%)
Strength of materials (7%)
Material properties (7%)
Fluid mechanics (7%)
Electricity and magnetism (9%)
Thermodynamics (7%)

The four-hour afternoon exam is comprised of 60 two-point questions and covers one of seven engineering disciplines (electrical, mechanical, civil, industrial, chemical, environmental, general) chosen by you.

The FE exam can be taken prior to graduation in engineering, ideally sometime in your senior year, or soon after you graduate. This is the time when you have the best command of engineering fundamentals. Once you have passed this exam and graduated, you are designated as an Intern-Engineer or Engineer-in-Training.

One note. Your required engineering curriculum may not include courses that cover all of the topics included on the FE exam, so you may need to take an EIT review course, do extensive self-study, or even elect to take an additional course or two (e.g., engineering mechanics, thermodynamics, material science) during your undergraduate years.

The Principles and Practice of Engineering Exam. After four years of acceptable experience as an Intern Engineer or Engineer-in-Training, you will be eligible to take the Principles and Practice of Engineering Exam (PE Exam). Offered in April and October, the PE

Exam is an eight-hour multiple-choice exam in a specific engineering discipline (civil, mechanical, electrical, chemical, industrial, etc.). The four-hour morning session consists of 40 questions covering the discipline broadly. The afternoon session consists of 40 questions in a subspecialty of the discipline selected by you.

PROFESSIONAL SOCIETIES

Each of the engineering disciplines described in Section 2.6 has a professional society that serves the technical and professional needs of engineers and engineering students in that discipline. These societies are usually organized on both national and local levels, and most support the establishment of student chapters on university campuses. The societies publish technical journals and magazines, organize technical conferences, sponsor short courses for professional development, develop codes and ethical standards, and oversee the accreditation of engineering programs in their discipline.

The benefits of getting actively involved in the student chapter corresponding to your engineering discipline will be discussed in Chapter 7. In the meantime, you can gain valuable information about the various engineering disciplines by exploring the web pages listed below.

Professional Society	Web Site
American Academy of Environmental Engineers (AAEE)	http://www.aaee.net
American Institute of Aeronautics and Astronautics (AIAA)	http://www.aiaa.org
American Institute of Chemical Engineers (AIChE)	http://www.aiche.org
American Nuclear Society (ANS)	http://www.ans.org
American Society of Agricultural and Biological Engineers (ASABE)	http://www.asabe.org
American Society of Civil Engineers (ASCE)	http://www.asce.org
American Society of Heating, Refrigerating, and Air-Conditioning Engineers (ASHRAE)	http://www.ashrae.org

American Society of Mechanical Engineers (ASME)	http://www.asme.org
Biomedical Engineering Society (BMES)	http://www.bmes.org
Institute of Electrical and Electronics Engineers (IEEE)	http://www.ieee.org/portal/site
Institute of Industrial Engineers (IIE)	http://www.iienet2.org
Society of Automotive Engineers (SAE)	http://www.sae.org
Society of Manufacturing Engineers (SME)	http://www.sme.org
Society for Mining, Metallurgy, and Exploration (SME-AIME)	http://www.smenet.org
Society of Naval Architects and Marine Engineers (SNAME)	http://www.sname.org
Society of Petroleum Engineers (SPE)	http://www.spe.org
The Minerals, Metals & Materials Society (TMS)	http://www.tms.org

SUMMARY

In this chapter you were introduced to the engineering profession—past, present, and future. You were encouraged to take every opportunity to learn as much as you can about engineering. This will be a lifelong process, but it has already begun.

First, we helped you develop an articulate answer to a question you are likely to be asked frequently: "What is engineering?" You learned that, at its core, engineering is the process of developing a product or process that meets a customer need or perceived opportunity.

Next, we gave you a view of the past by presenting the 20 Greatest Engineering Achievements of the 20th Century. Reading about these not only provided an interesting retrospective of the engineering field; hopefully, it also served as an incentive to you as a new engineering student. For the achievements of the 21st century are bound to be even more spectacular than the accomplishments of the 20th century. And <u>you</u> may very well be responsible for one of the "Greatest Engineering

Achievements of the 21st Century." In any event, whether you look backward to the past or forward to the future, you can see what an important and exciting profession you will be joining.

We then discussed the many rewards and opportunities that will be yours if you are successful in graduating in engineering. By developing individual "top ten" lists of these rewards and opportunities, starting with a discussion of "Ray's Top Ten" list, you should have a clear picture of how an engineering degree will greatly enhance the quality of your life—as well as the lives of others.

Next, to flesh out your understanding of the engineering profession, you saw how engineers can be categorized according to their academic discipline and job function, each of which we studied in some detail.

You then learned about the employment opportunities that will await you upon graduation. The North American Industry Classification System (NAICS) was used to give you a feel for the enormous economic engine that your engineering education is preparing you to be part of. We paid particular attention to the industry sectors that employ the largest numbers of engineers, and discussed the technical fields that are expected to grow rapidly in the future. You may want to begin preparing yourself today for a career in one of these "hot" fields of the future.

Finally, you learned that engineering is a profession that you will enter when you graduate. We discussed the requirements that define a "profession," including the rights and privileges that come with responsibilities and obligations. Among the rights you will be accorded is to become licensed as a registered Professional Engineer (P.E.). You will also have the opportunity to participate in the engineering society appropriate to your academic discipline, both while you are in school and throughout your engineering career.

REFERENCES

1. "2005 ABET Accreditation Yearbook," Accreditation Board for Engineering and Technology, Inc., Baltimore, MD, 2005.

2. Constable, George and Somerville, Bob, *A Century of Innovation: Twenty Engineering Achievements That Transformed Our Lives*, Joseph Henry Press, Washington, DC, 2003.

3. Collins, James C. and Porras, Jerry I., "A Theory of Evolution," *Audacity: The Magazine of Business Experience*, Vol. 4, No. 2, Winter, 1996.

4. Ulrich, Karl T. and Eppinger, Steven D., *Product Design and Development, Third Edition*, McGraw-Hill/Irwin, 2004.

5. Kyle, Chester R., *Racing with the Sun: The 1990 World Solar Challenge,* Society of Automotive Engineers, Warrendale, PA, 1991.

6. King, Richard and King, Melissa, *Sunracing,* Human Resource Development Press, Amherst, MA, 1993.

7. Carroll, Douglas R., *The Winning Solar Car: A Design Guide for Solar Car Teams*, SAE International, October, 2003.

8. Landels, John G., *Engineering in the Ancient World*, University of California Press, 1981.

9. "May 2005 National Occupational Employment and Wage Estimates - United States," U. S. Department of Labor, Bureau of Labor Statistics. http://www.bls.gov/oes/current/oes_nat.htm

10. "Digest of Education Statistics Tables and Figures 2005," U.S. Department of Education, National Center for Education Statistics. http://nces.ed.gov/programs/digest/d05/lt3.asp#19

11. "U.S. Job Satisfaction Keeps Falling," The Conference Board, February 2005. http://www.conference-board.org/utilities/pressDetail.cfm?press_ID=2582

12. "Building A Better Brain," *Life Magazine*, p. 62, July, 1994.

13. "Salary Survey: A Study of 2005-2006 Beginning Offers," Volume 45, Issue 4, National Association of Colleges and Employers, 62 Highland Avenue, Bethlehem, PA 18017, Fall, 2006 (http://www.naceweb.org).

14. "List of Billionaires (2006)," Forbes Magazine, 2006. http://en.wikipedia.org/wiki/List_of_billionaires

15. Lumsdaine, Edward, Lumsdaine, Monika, and Shelnutt, J. William, *Creative Problem Solving and Engineering Design*, McGraw-Hill, New York, 1999.

16. Gibbons, Michael, "The Year in Numbers," American Society for Engineering Education, Washington, DC, 2005. http://www.asee.org/publications/profiles/upload/2005ProfileEng.pdf

17. "Computer Engineering 2004: Curriculum Guidelines for Undergraduate Degree Programs in Computer Engineering, " IEEE Computer Society, 2004.

18. Landis, R. B., "An Academic Career: It Could Be for You," American Society for Engineering Education, Washington, D.C., 1989.

19. "Science and Engineering Indicators 2006," Volume 2, Table 3-9, National Science Foundation, SESTAT Database (Available on line at: http://sestat.nsf.gov)

20. "North American Industry Classification System (NAICS) - United States, 2002," U.S. Census Bureau, 2002. (Available at: http://www.census.gov/epcd/www/naics.html)

21. Friedman, Thomas L., *The World is Flat: A Brief History of the Twenty-First Century*, Farar, Straus and Giroux, New York, 2005.

22. "The Long View," National Science Foundation Publication 93-154, Arlington, VA, 1993.

23. "Strategic Planning Overview: Strategic Directions for Engineering Research, Innovation, and Education," National Science Foundation, Directorate for Engineering, June, 2005.

24. "The Engineer of 2020: Visions of Engineering in the New Century," National Academy of Engineering, Washington, DC, 2004.

25. Beakley, G. C., Evans, D. L., and Keats, J. B., *Engineering: An Introduction to a Creative Profession,* 5th Edition, Macmillan Publishing Company, New York, NY, 1986.

PROBLEMS

1. Review the definitions of engineering presented in Appendix A. Combine the best ideas from these definitions, write out your own definition of "engineering," and memorize it. Ask people you come in contact with whether they know what engineering is. If they say, "No," then recite your definition to them.

2. Review the National Engineers Week web page:

   http://www.eweek.org

   and answer the following questions:

   a. Who are the sponsors of National Engineers Week?
   b. What is the purpose of National Engineers Week?
   c. What are some of the major activities scheduled for the next National Engineers Week celebration?

d. What are some of the products available to help promote National Engineers Week?

3. Write a one-page paper about the influences (teachers, parents, TV, etc.) that led you to choose engineering as your major.
4. Pick one of the engineering guidance web sites listed at the end of Section 2.1. Explore the web site to learn as much about engineering as you can from it. Write a one-page paper summarizing what you learned.
5. Write a list of specifications for a motorized wheel chair that could be used on a sandy beach. Include performance specifications, economic specifications, and scheduling specifications.
6. Review the list of needed products at the end of Section 2.3. Add five additional needed products that you think would sell if developed into actual products.
7. Pick one of the items from Problem 6 and write a set of design specifications for the proposed product.
8. Pick one of the 20 Greatest Engineering Achievements of the 20th Century. Write a one-page paper describing the impact of that engineering achievement on the quality of your life.
9. Create a list of activities you can do that will increase your understanding of engineering careers. Develop a plan for implementing three of these activities.
10. Add ten or more additional items to the list of rewards and opportunities of an engineering career presented in Section 2.5. Pick your top ten from the total list and rank them in order of importance.
11. Write a three-page paper on "Why I Want to be an Engineer" by expanding on your top four items from Problem 10 and explaining why each is important to you personally.
12. Have you ever had a job you didn't like? Describe the job. What didn't you like about it? If that job played any role in your subsequent decision to major in engineering, explain what that role was.
13. Read a biography of one of the famous people listed in Section 2.5 who were educated as an engineer. Make a list of the ways their engineering education supported their achievements.
14. Write down five non-engineering careers (e.g., politician, entrepreneur, movie director, etc.) that you might be interested in.

Discuss how obtaining your B.S. degree in engineering could help you pursue each of those careers.

15. What is the most challenging problem you have ever tackled in your life? Were you able to succeed at solving the problem? Did you enjoy the experience? Write a two-page essay that addresses these questions.

16. Answer the following questions related to making money:

    a. What is the legal "minimum wage" (per hour) in the U.S.?

    b. What is the highest hourly wage you have ever made?

    c. What hourly wage would correspond to the average starting annual salary for engineering graduates in 2005/06 ($51,465)?

    d. What is the hourly wage of an engineering executive making $250,000 a year?

17. As indicated in Section 2.5, engineering graduates make up only 4.5 percent of all college graduates. Go to your career center and find out how many employers interview on campus annually. What percentage of those employers interview engineering majors only? What percentage interview business majors only? What percentage interview all other majors? What is the significance of your findings?

18. Find out how the following things work (if you don't already know):

    a. Fuel cell
    b. Radar gun
    c. Microwave oven
    d. Solar cell
    e. Digital display

    Prepare a three-minute oral presentation about one of the items that you will give at the next meeting of your Intro to Engineering course.

19. Go to the National Engineers Week web page (http://www.eweek.org) and click on "The Creative Engineer." There you will find eight elements of creativity:

    challenging connecting
    visualizing collaborating
    harmonizing improvising
    reorienting synthesizing

    Pick one of these elements. Look up the definition of the term in the dictionary, study the example on the National Engineers Week web page, and conduct further research on the element. Write a one-page paper explaining why this "element of creativity" is important in engineering work.

20. Go to Professor John Lienhard's *The Engines of Our Ingenuity* web page: http://www.uh.edu/engines. Pick three of the more than 2,150 episodes you will find there. Study those three. Write a two-page paper on why you picked the ones you did and what you learned from studying them.

21. In Section 2.6, you learned that engineering disciplines can be divided into two categories: (1) the six largest disciplines (electrical, mechanical, civil, computer, chemical, and industrial); (2) a much larger number of smaller, more specialized disciplines. Make a list of the advantages and disadvantages of selecting your major in one or the other of these categories.

22. Which of the engineering disciplines listed in Section 2.6 are offered by your university? Find out how many students graduate annually from your university in each of these disciplines.

23. Pick one of the engineering disciplines listed in Section 2.6. Visit the web page of the professional society corresponding to that discipline and take note of any information that applies specifically to engineering students. Share what you learned with your classmates in your next Introduction to Engineering class.

24. Pick one of the engineering disciplines listed in Section 2.6. Write a three-page paper describing that discipline.

25. Pick one of the technical divisions or societies of either the American Society of Mechanical Engineers (ASME), the Institute of Electrical and Electronics Engineers (IEEE), or the American Society of Civil Engineers (ASCE) listed in Section 2.6 that you would like to know more about. Research the division or society and write a one-page paper describing it.

26. Which of the civil engineering specialties described in Section 2.6 would provide you the greatest opportunity to benefit society? Why?

27. Go to the U.S. Bureau of Labor Statistics web-based "Occupational Outlook Handbook" at: http://www.bls.gov/oco/ocos027.htm. Study the information there to learn as much as you can about "engineers." What does it say about the job outlook for "engineers"?

28. Go to the American Institute of Chemical Engineers (AIChE) "CareerEngineer" web page: http://www.aiche.org/CareersEducation. Click on "Find a Job," and read about the ones listed there. Which one appeals to you the most? Prepare a two-minute talk describing its appeal to your Introduction to Engineering classmates.

29. Interview a practicing engineer. Find out the answers to the following questions:
    a. What engineering discipline did he or she graduate in?
    b. To what extent do the knowledge and principles of that discipline apply in his or her current job?
    c. What industry sector does he or she work in?
    d. What percentage of his or her time is spent in the various engineering job functions (design, test, development, management, etc.)?

30. Develop a list of attributes that would be desirable for each of the engineering job functions described in Section 2.7. Which of these job functions appeals most to you? Be ready to explain your reasons in a class discussion when your Intro to Engineering course next meets.

31. Familiarize yourself with the NAICS system by doing the following exercise. Begin by accessing the Internet. Then proceed as directed by the steps below:
    a. Go to: http://www.census.gov
    b. Click on "NAICS"
    c. Click on "NAICS Search"
    d. Enter keyword "NAICS 334"
    e. Browse through all the products listed under NAICS 3345
    f. Find the products listed under NAICS 334510 and print them out

32. Pick one of the products listed under NAICS 334510 from Problem 31 and research what companies manufacture that product. Contact that company and investigate how they use engineers in the design, manufacture, test, and marketing of that product. Write a summary of what you learned.

33. Learn about the "Engineering Services" industries by following the steps outlined in Problem 31 and entering the keyword "NAICS

541330." How many entries did you find? Would you be interested in working in this industry? Why or why not?

34. Obtain a list of employers that conduct on-campus interviews of engineering graduates through your career center. Try to identify which industry sector each employer belongs in, based on those listed in Section 2.8. Do some of the employers fit into more than one industry sector?

35. Identify the two or three engineering disciplines that you think would be most closely associated with each of the eight "Service" Economic Sectors and six "Manufacturing" Economic Subsectors described in Section 2.8.

36. Make a list of ten products that would be manufactured by each of the six "Manufacturing" Economic Subsectors listed in Section 2.8.

37. Pick one of the important fields for the future described in Section 2.9. After researching this field, write a three-page paper that first describes the field in detail and then discusses future employment opportunities that the field will offer.

38. Explain how each of the "Major Events and Changes" listed in Section 2.9 will impact your future. What effect will each have on the engineering job market? (Will it increase or decrease the number of jobs? In which disciplines? Will it change the nature of current jobs?)

39. Make a list of 1) the rights and privileges; and 2) the responsibilities and obligations you will have when you join the engineering profession.

40. Obtain information about the process of becoming a registered Professional Engineer in your state. How do the requirements and procedures differ from those presented in Section 2.10? What engineering disciplines are licensed in your state?

41. Set a personal goal of passing the Fundamentals of Engineering Exam (FE) before or soon after you graduate. Develop a set of strategies that will ensure that you are well prepared to pass the exam.

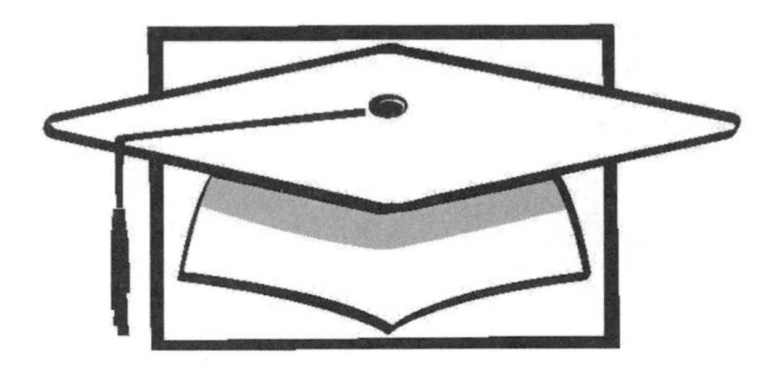

CHAPTER 3
Understanding the Teaching/Learning Process

Education is not preparation for life;
Education is life itself.

John Dewey

INTRODUCTION

In Chapter 1, we identified "approach" as a key factor that will lead you to success in your engineering studies. We linked "approach" with "effort," another key factor, explaining that the successful student is one who works both *hard* and *smart*. In this and the two subsequent chapters, we will focus on what it means to work "smart."

In this chapter, we will provide an overview of the "teaching/ learning" process. Understanding this process will help you take full advantage of both the teaching part and the learning part.

We begin by discussing the learning process. We define learning and describe its components—receiving and processing new knowledge; and demonstrating mastery of that knowledge. You will have the opportunity to find out how you prefer to both receive and process new knowledge—insights that will be useful to you in designing your own learning process. We will also provide you with the characteristics of "expert" learners and encourage you to continuously improve your learning skills by observing your learning process and making changes based on those observations.

We conclude the section on learning with the important perspective that learning is a reinforcement process. This is an overarching principle that you should take advantage of at every opportunity.

Next, we provide a brief overview of the teaching component of the "teaching/learning" process. We point out the ways teaching is delivered, including various teaching styles used by your professors. Being aware of how you are being taught can help you create your learning experience.

Then we point out some common mistakes students make as they transition from high school to engineering study. An overarching (and costly) mistake is to assume that the same strategies and approaches that worked in high school will work here. We also provide an indication of approaches and strategies for avoiding these mistakes.

We conclude the chapter by providing a perspective on the subject of seeking help. Failing to utilize available resources—particularly professors and fellow students—is a serious mistake that some students make. "Standing on the shoulders" of others is fundamental to the very concept of an education.

3.1 What is Learning?

Although you've been an active participant in the learning process for many years, it is unlikely that you have had any formal training in it. Becoming an expert learner requires not only that you devote time and energy to learning, but that you devote time and energy into learning *how* to learn.

Learning, broadly defined, is the process of acquiring:

- New knowledge and intellectual skills (cognitive learning)
- New manual or physical skills (psychomotor learning)
- New emotional responses, attitudes, and values (affective learning)

Cognitive Learning

Cognitive learning is demonstrated by knowledge recall and higher level intellectual skills. Since acquiring new knowledge and intellectual skills will be the primary focus of your engineering education, it's useful to explore the process in some detail.

Bloom's Taxonomy as modified by Anderson [1] identified six levels of intellectual skills within the cognitive domain. At the lowest level is the simple recall or recognition of facts. The highest level is classified as creating. In between are increasingly more complex and abstract mental levels. The following are brief definitions of each level of intellectual skill, along with examples that represent activity for each level.

Intellectual Skill	Definition	Examples of Intellectual Activity
Remembering	Retrieving, recognizing, and recalling relevant knowledge from long-term memory	arranging, defining (from memory), duplicating, labeling, listing, memorizing, naming, ordering, recognizing, relating, recalling, repeating, reproducing, stating
Understanding	Constructing meaning from oral, written, and graphic messages	classifying, describing, defining (in your own words), discussing, explaining, expressing, identifying, indicating, locating, recognizing, reporting, restating, reviewing, selecting, translating
Applying	Carrying out or using a procedure through executing or implementing	applying, choosing, demonstrating, dramatizing, employing, illustrating, interpreting, operating, practicing, scheduling, sketching, solving, using, writing
Analyzing	Breaking material into constituent parts; determining how the parts relate to one another and to an overall structure	analyzing, appraising, deriving, calculating, categorizing, comparing, contrasting, criticizing, differentiating, testing, discriminating, distinguishing, examining, experimenting, predicting, questioning
Evaluating	Making judgments based on criteria and standards through checking and critiquing	appraising, arguing, assessing, attaching, choosing, comparing, contrasting, critiquing, defending, judging, predicting, rating, selecting, supporting, evaluating
Creating	Putting elements together to form a coherent or functional whole; reorganizing elements into a new pattern or structure	arranging, assembling, collecting, composing, constructing, creating, designing, developing, formulating, managing, organizing, planning, preparing, proposing, setting up, writing

In high school, you primarily worked at the first two levels of demonstrating subject mastery—remembering and understanding. Often you could earn good grades by memorizing material and successfully repeating it back. In your university education, the expectations are higher. From the start, you will be expected to think at levels 3 and 4—applying and analyzing—mastering concepts and applying them to solve new problems. Your performance will be evaluated on your ability to extend what you learn in one context to new contexts. In your first two years, you may be given some problems that require you to think at levels 5 and 6—evaluating and creating, but the strongest focus on these levels of thinking will come in your junior and senior years.

The importance of acquiring higher intellectual skills is underscored in an excellent book by John Bransford titled *How People Learn* [2].

> All learning involves "transfer"—defined as the ability to extend what has been learned in one context to new contexts—from previous experiences. Educators hope that students will transfer learning from one problem to another within a course, from one school year to another, between school and home, and from school to the workplace. Transfer is affected by the degree to which people learn with understanding rather than merely memorize sets of facts or follow a fixed set of procedures.

I can assure you that mastering the learning skills presented in this chapter and the next two chapters will aid you in this process of "kicking it up a notch" from what was expected of you in high school.

PSYCHOMOTOR LEARNING

Psychomotor learning is demonstrated by physical skills—coordination, dexterity, manipulation, grace, strength, and speed, for example; actions that demonstrate the fine motor skills such as use of precision instruments or tools; or actions that demonstrate gross motor skills such as dance or athletic performance.

Within engineering education examples of learning in this domain might include sketching, computer keyboard skills, machine tool operation, and certain laboratory skills.

AFFECTIVE LEARNING

Affective learning is demonstrated by behaviors indicating attitudes of awareness, interest, attention, concern, and responsibility, ability to listen and respond in interactions with others, and ability to demonstrate those attitudinal characteristics or values which are appropriate to the test situation and the field of study. This domain relates to emotions, attitudes, appreciations, and values, such as enjoying, conserving, respecting, and supporting.

Many of the topics in this text fit into the affective learning category. Indeed, one of the primary purposes of this book is to guide your personal development in many of the areas listed above. This development will continue throughout your engineering education, even though it might not be addressed explicitly in your required coursework. For example, through your participation in informal study groups and laboratory and design project groups, you will grow in your ability to "listen and respond in interactions with others." However, you might have to elect to take a course in *Interpersonal Communication* to gain explicit training in this area.

3.2 How Do We Learn?

The process of learning new knowledge and intellectual skills (cognitive learning) can be thought of in two steps: 1) receiving new knowledge; and 2) processing that new knowledge.

Receiving New Knowledge

There are two key aspects to the way you receive new knowledge:

- the type of information you prefer
- the sensory channel through which external information is most effectively perceived

Research has shown that learners differ with regard to their preferences for receiving new knowledge in each of these two areas.

I. What Type of Information Do You Prefer?

Sensing learners. Sensing learners focus on things that can be seen, heard, or touched. They like facts and data, the real world, and above all, relevance. They are patient with details and enjoy solving problems by well-established methods.

Intuitive learners. Intuitive learners are dreamers. They prefer ideas, possibilities, theories, and abstractions. They look for meanings, prefer variety, and dislike repetition. They tend to dislike "plug and chug" courses and to be impatient with detailed work.

II. Through What Sensory Channel Do You Perceive External Information Most Effectively?

Visual learners. Visual learners learn more effectively through the use of pictures, diagrams, flow charts, graphs, sketches, films, and demonstrations.

Verbal learners. Verbal learners respond more to the written or spoken word. They like to read about things or hear explanations from an expert.

Processing New Knowledge

There are two important aspects to processing new knowledge:

- the way you prefer to process information
- the way you progress toward understanding

Again, research has shown that learners differ with regard to their preferences in these two areas.

III. The Way You Prefer to Process New Information

Active learners. Active learners tend to process information while doing something active with it. Consequently, active learners think out loud, try things out, and prefer group work. Sitting through lectures is particularly hard for active learners.

Reflective learners. Reflective learners prefer to think about information quietly first. They want to understand or think things through before attempting to do anything themselves. They tend to prefer working alone.

IV. The Way You Progress Toward Understanding

Sequential learners. Sequential learners prefer linear steps, with each step following logically from the previous one. They tend to follow logical stepwise paths in finding solutions.

Global learners. Global learners tend to learn in large jumps, absorbing material almost randomly without seeing connections, and then suddenly "getting it." They prefer to see the "big picture" and then fill in the details.

Index of Learning Styles Questionnaire

You may already have a sense of your preferred learning styles. If you'd like to get a more definitive indication, however, I encourage you to complete the *Index of Learning Styles Questionnaire* developed by Barbara Solomon and Richard Felder at North Carolina State University. This questionnaire can be completed on-line at:

http://www.engr.ncsu.edu/learningstyles/ilsweb.html

At that web site, you will be asked to choose one of two preferences for 44 items that cover four areas (two areas related to receiving information; two areas related to processing information).

You will immediately receive the scored results indicating that you have either a strong preference, moderate preference, or are fairly well balanced on each of the four scales.

As you'll learn later in this chapter, if your preferred ways of receiving new knowledge are ***verbal*** and ***intuitive*** and your preferred ways of processing new knowledge are ***sequential*** and ***reflective***, you are in step with the ways you are most likely being taught (i.e., traditional lecture format). But you can almost certainly benefit by adopting the full spectrum of learning styles in your studies (sensory as well as intuitive; visual as well as verbal; active as well as reflective; global as well as sequential).

If your preferred ways of receiving new knowledge are ***visual*** and ***sensory*** and your preferred ways of processing new knowledge are ***global*** and ***active***, you may find a mismatch between the way you prefer to learn and the way you are being taught. The good news is that you are primarily responsible for creating your learning experience. So just make sure that the way you study is compatible with your preferred learning styles. Some general advice from Richard Felder as to how to do this can be found at:

http://www.ncsu.edu/felder-public/ILSdir/styles.htm

Keep in mind, however, that:

All learners benefit from using learning styles on both sides of all four dimensions

Doing things that are compatible with your style can help you overcome mismatches with the dominant style of your teachers. Doing things that are on the opposite side of your learning style preference will give you a perspective on the material that you might not normally get and help you develop skills that will enhance your professional success—skills that you might not develop if you only followed your natural inclinations.

3.3 Metacognition – The Key to Improving Your Learning Process

Famous Swiss psychologist Carl Jung put forth the concept of the "observing ego." This is the part of you that observes your "self." It observes what you do. It observes how you think. And it observes how you feel. And through this process of "watching," it feeds back information (assessment) and enables you to make changes. Some examples of this process are shown below.

Activity	Observation	Feedback	Change
Running a 10K race	I'm running too fast.	I'd better slow down.	Slow down pace
Falling asleep while studying	I'm not going to get my homework done.	A break will help.	Take a walk around the block
Doing home-work problems	I can't do Problem #5.	I need to seek help.	Go see instructor during office hours
Attending math class	I'm not following what's being presented.	I could get clarification by asking a question.	Raise hand and ask question

The process of improving your learning process by observing it, feeding back what you observe, and making changes is called *metacognition*. The word is not as important as the process itself. There are three important steps in doing this: 1) planning your learning; 2) monitoring your learning; and 3) evaluating your learning. I would encourage you to begin the process of watching your learning. The easiest way to do this is to start asking yourself questions.

1. When ***planning your learning,*** ask questions like:

- What in my prior knowledge or experience will help me with this particular task?
- What should I do first?
- Why am I doing this task?
- How much time do I have to complete this task?

2. To ***monitor your learning***, ask questions like:

- How am I doing?
- Am I on the right track?
- How should I proceed?
- What information is important to remember?
- Should I move in a different direction?
- Should I adjust the pace based on the difficulty (too difficult – slow down; too easy – speed up)
- What do I need to do if I do not understand?

3. To ***evaluate your learning,*** ask questions like:

- How well did I do?
- Did my particular course of action produce more or less than I had expected?
- What could I have done differently?
- How might I apply this line of thinking to other problems?
- Do I need to go back through the task to fill in any "blanks" in my understanding?

REFLECTION

Reflect on the following list of characteristics of "expert" learners:

Control the learning process rather than become a victim of it
Are active, not passive, in their approach to learning
Are motivated (e.g., enjoy learning, have short-term and long-term goals, etc.)
Are disciplined (i.e., have learned good habits and use them consistently)
Are more aware of themselves as learners (e.g., know their own strengths and weaknesses)
Initiate opportunities to learn
Set specific learning goals for themselves
Have a larger repertoire of learning strategies from which to choose
Know not only *what* to learn, but *how* to learn
Plan their approach to learning
Monitor their learning while it's happening
Are more adaptive because they *do* self-monitor while learning
Reflect more upon their own learning
Evaluate the effectiveness of learning approaches and strategies
Are more sensitive to the demands of specific academic tasks
Use learning strategies selectively
Tend to attribute failures to correctable causes
Tend to attribute successes to personal competence

Rate yourself on a scale of 0-10 on each of these items. Are there areas you need to improve in? Are you willing to work at doing so? Pick two or three areas you would like to improve in and for each area come up with three specific things you can do.

I hope I have persuaded you of the benefits of "observing" your learning process with the goal of becoming an "expert" learner.

3.4 LEARNING IS A REINFORCEMENT PROCESS

There is one more overarching principle that should influence every aspect of your participation in the "teaching/learning" process. It is that:

Learning is a reinforcement process

Learning comes from repeated exposure to the subject material—the more the better. A critical part of the learning process is what we call "reinforcement."

An Example: The Study of Mechanics

The way in which we learn the subject of mechanics, the study of forces and motion, can illustrate the importance of reinforcement in the learning process.

Our first exposure to mechanics may have come in high school physics. Next, we study a whole semester of mechanics in our freshman physics course. In the sophomore year, we may have a course in statics and, in the junior year, a course in engineering dynamics. If we are interested, we can take several senior-level courses and, for a thorough understanding of mechanics, we could pursue graduate study—a master's or even Ph.D. degree.

Even then, if we were to begin to teach mechanics, we would find areas where we were not completely clear, and probably only after a number of years of teaching would we feel that we were even approaching total mastery of the subject.

The point of this example is not to discourage you, but rather to encourage you to take advantage of every opportunity to reinforce your learning. Even for the brightest person, learning is a slow process that occurs over time and relies on repeated reinforcements. By knowing this, hopefully you will not fall into the common trap of thinking that you can "cram" in the material the night before a test.

The educational system is structured to give you the opportunity to reinforce the subject matter many times within a semester or quarter. Here are some examples of the reinforcement process at its best.

When	What To Do
Before class	Prepare for the lecture by reviewing notes, reading text, attempting a few problems, formulating some questions

During class	Attend lecture, concentrate intently, take detailed notes, ask questions
After class, but before next class meeting	Review and annotate notes, reread text; work assigned problems, work extra problems, meet with a study partner or study group to go over material and problems
In preparation for test or exam	Review notes; review text, rework problems, meet with a study partner or study group to go over material and problems
In preparation for final exam	Review notes, reread text, rework problems, meet with a study partner or study group to go over material and problems

This systematic approach to learning that involves repetition, review, and reinforcement will carry you a long way toward becoming an "expert" learner. Once you master the specific skills presented in the next two chapters as well, you'll be an "expert" learner for sure.

"I forgot to make a back-up copy of my brain, so everything I learned last semester was lost."

If you keep the principle that learning is a "reinforcement process" in mind as you design your learning process, you won't find yourself in the situation shown above.

3.5 Understanding the Teaching Part of the Teaching/Learning Process

The "teaching" part of the teaching/learning process is primarily achieved by the following well-known ***teaching modes***:

- Large lectures, in which one professor lectures to 100 to 300 or more students
- Small lectures, in which one professor lectures to 25 to 30 students
- Recitations, in which a teaching assistant reviews the material and solves problems for small groups of ten to 15 students
- One-on-one tutoring, in which a tutor works with one student

Despite their obvious differences, all four teaching modes have features in common. Each involves a person who is knowledgeable about a subject (an "expert," if you will) communicating what he or she knows to a less knowledgeable person (the student). Generally, most of the communication is one-way—i.e., from the teacher to the student. And most importantly, students learn relatively little from participating in any of these modes.

That last feature should alarm you, or at least cause you to question how I could make such a provocative statement. Here's how. Imagine that you're in an engineering course, and your professor introduces a new principle. You go to the lecture, recitation, and tutoring sessions, but you don't do anything outside of those activities. Then you are given an exam on the principle. What score would you expect to make?

The limited effect of these teaching modes—especially the lecture format—becomes quite apparent if you envision the process as one educator has aptly described it:

> ***The information passes from the notes of the professor to the notes of the student without passing through the mind of either one.***

TEACHING STYLES

Even within the general lecture format, there are variations in the way the teaching is done. In reading Section 3.2, you should have discovered your preferred learning styles. Your effectiveness in creating your learning experience will be related to your understanding not only of the way you best learn, but also seeing it in the context of how you are being taught. The following is an overview of the various dimensions of teaching styles.

Dr. Richard Felder of North Carolina State University, a world-renowned expert on teaching and learning styles in engineering education [3], examines five aspects of the teaching process and describes two possibilities for the approach your professors might take for each of these categories. The teaching styles most prevalent in math/science/engineering courses (abstract, verbal, deductive, passive, and sequential) are highlighted in bold type.

1) **What type of information is emphasized?**

   Concrete – Facts, data, observable phenomena

   Abstract – Principles, concepts, theories, mathematical models

2) **What mode of presentation is stressed?**

   Visual – Pictures, diagrams, films, demonstrations

   Verbal – Spoken words, written words

3) **How is the presentation organized?**

   Deductive – Start with fundamentals; proceed to applications

   Inductive – Start with applications; proceed to fundamentals

4) **What mode of student participation is facilitated?**

   Active – Student involved (talking, moving, reflecting, solving problems)

   Passive – Student as spectator (watch, listen)

5) **What type of perspective is provided on the information presented?**

   Sequential – Step by step progression

   Global – Context and relevance are provided

REFLECTION

Think about the classes you are taking in terms of each of the five categories above. Do your instructors predominantly use the teaching styles highlighted in bold type? Or do they use the other one? Or do they use some combination of the two? Pick the teaching style under each of the five categories above that you would learn best from and place an asterisk next to it. How many match the one most common in your courses? How many differ? What meaning does this have for you?

By now, I hope you're asking yourself some very good questions, like:

- What value is it to me to understand how my professors teach?
- If the way my professors teach differs from the way I prefer to learn, does that mean I can't learn the material and I am perhaps in the wrong major?
- If some students prefer to be taught in one way and other students prefer to be taught in another way, why wouldn't the professor use both ways?

These are complex and interesting questions and I would encourage you to seek out the answers. Here are some of my thoughts for starters.

BENEFITS OF UNDERSTANDING HOW PROFESSORS TEACH. There are several benefits to you in understanding the different ways of teaching. Perhaps primary among these is that the knowledge will guide you in designing your learning process.

But equally important is that you will be doing lots of teaching yourself. When I was dean of engineering, from time to time someone would ask me: "Do you do any teaching?" Even though I knew they really meant *Was I teaching any courses?* and I wasn't, I would respond "I'm always teaching." Throughout your time as a student and during your professional career, you will constantly be teaching others what you know—in formal and informal meetings; through your written communications; and in almost everything you do. Understanding the different ways of transmitting information and knowledge will be an extremely useful skill in all aspects of your life and career. And who knows? Maybe someday you'll decide to become an engineering professor.

WHAT IF THE WAY I PREFER TO LEARN DIFFERS FROM THE WAY I AM TAUGHT? Rather than view this as a problem, I suggest that you view it as an opportunity. Just because you prefer to be taught in one of two possible ways doesn't mean that you can't learn the other way. During your learning process, you can learn even more by translating what you were taught into the way you prefer to learn. For example, if you need to know the context and relevance (global perspective) for what you are being taught and your professor doesn't provide it, it's likely that you will learn even more if you develop it than you would have if the professor had provided it.

Another point is that there is a difference between preference and competence. You may like doing something, but not be good at it. For example, I love to sing, but I can't carry a tune. Conversely, you may be good at something, but not like doing it. I'm sure I'd be good at accounting, but I wouldn't want to do it for a living. Strive to improve your ability to get the most out of all teaching styles—the ones you prefer and the ones you don't. You might prefer to study by yourself but find that you are very good at studying collaboratively with other students. As a student and as a professional you'll be learning from and teaching people who have different preferences from yours. Your effectiveness will depend on your "agility" in trying different ways of getting your points across.

The bottom line is that you will benefit by developing your skills on both sides of all four learning style dimensions. Good engineers are observant and methodical, practical, willing to check calculations and replicate experiments over and over to be sure they're right (sensing skills). They also have to be innovative, to deal with theories and models, and to think deeply about the meaning of their observations and results (intuitive skills). Good engineers have to deal with both visual and verbal information, reflect on things and take action; be aware of the big picture and proceed in a stepwise manner.

REFLECTION

Think about your ability to do various things. Are there things you like to do, but don't do well? Are there things that you do very well, but don't like doing? Or do you generally only like doing the things you do well? How can understanding the difference between preferences and competence benefit you in your education?

So there's no need for you to drop out of school or to change majors. You can learn to learn no matter how you are being taught. And besides there's a place in the engineering profession for individuals with different preferences for the way they are taught. Think about the various engineering job functions described in Chapter 2—analysis, design, test, development, sales, research, management, consulting, teaching. Although one individual might be best suited for one (e.g., someone who prefers to be taught visually might make a good design engineer) and another individual best suited for another (e.g., someone who prefers to be taught verbally might make a good analytical engineer), all engineering job functions require individuals who can both "learn" and "teach" in a variety of styles.

WHY DON'T YOUR PROFESSORS USE A VARIETY OF TEACHING STYLES? Many do. And in time, more and more will. There is an impetus for change within engineering education. *Interactive lectures, problem-based learning, inquiry-guided learning,* and *just-in-time teaching* are a few examples of teaching methods that are gaining acceptance within engineering education. These methods bring more active student involvement and context and relevance into the classroom.

But change is slow. Formal training in teaching methods is not a required part of the process of becoming a math, science, or engineering professor. In the absence of such training, most professors tend to teach the way they were taught. And so the most prevalent teaching styles of the past (abstract, verbal, deductive, passive, sequential) tend to be propagated into the future.

Whatever you do, don't use the way your professor teaches as an excuse for not learning. If you believe you are having difficulty because of the way you are being taught, you might speak to your professor and suggest ways in which he or she could help you get more out of the lectures. Suggestions might include working more problems, providing the "big-picture" for the concepts being covered, or recommending additional resources that would aid you in your learning process. One caution. Do this politely and in a constructive tone. No one likes to think he or she is being criticized.

3.6 MISTAKES STUDENTS MAKE

As we discussed in Chapter 1, engineering study is challenging, but you have the ability to do it. What's critical for your success is to avoid many of the mistakes that students make early on. If you do, you're likely to be one of the many who succeed. If you don't, you'll increase the chance you'll be among those who don't succeed.

An overarching mistake is to assume that your college engineering studies will be like high school and that the same strategies and approaches that worked there will work here. In reality, the faster pace and higher expectations for learning require new strategies and improved learning skills. Included with each mistake listed below is a general strategy for overcoming it. Taking advantage of these success strategies will require that you learn, practice, and refine the learning skills that are described in this chapter and in Chapters 4 and 5.

Mistakes Students Make	Strategies for Overcoming Them
Assume engineering study will be like high school	Work to understand and adjust to the differences between high school and engineering study
Program themselves for failure through too many commitments	Create a life situation that enables you to devote adequate time and energy to your studies
Spend little time on campus	Immerse yourself in the academic environment of the institution
Neglect studying	Schedule study time. Devote significant time and energy to studying.
Delay studying until test is announced	Master the material presented in each class prior to next class
Study 100% alone	Study collaboratively with other students
Come to each lecture unprepared	Review notes, read text, attempt problems prior to each lecture
Avoid professors	Interact regularly with professors outside the classroom
Cut classes and/or don't get the most out of lectures.	Attend classes and practice good listening skills. Ask questions in class.
Fail to take notes; or take notes but fail to use the notes properly in the learning process	Take effective notes and use a systematic learning methodology to study from notes
Skim over the material in an assigned chapter in a rush to get to the assigned homework problems	Use reading for comprehension methodology to understand the general concepts thoroughly before attempting problems
Fail to solve assigned problems. Don't approach problems using a systematic problem solving method	Solve not only assigned problems but extra problems; use systematic problem solving methods

Our purpose in this chapter and in Chapters 4 and 5 will be to guide you in developing important learning strategies and skills that will move you from the left hand column into the right hand column. If you are able to do this, the likelihood you will succeed in your engineering studies will be greatly enhanced.

Reflection

Review the list of "mistakes students make." Do any of these describe the way you are approaching your engineering studies? Do you think you would benefit by changing to the strategy described in the right hand column? Pick one or two items and make a commitment to the required changes.

3.7 Don't Be Hung Up on the Idea of Seeking Help

Do you feel that seeking help is a sign of weakness? That if you make it on your own you will get more out of your education? The idea that we can make it through life without help from others is simply not true.

The truth is that we all rely heavily on others to live, grow, and thrive. We come into this world totally dependent on our parents or guardians for our very survival. As we grow, most of what we learn we are taught by others—parents, family, teachers, peers. In school when we use a textbook, in engineering or just about any other discipline, we benefit from the many experts who have evolved the subject over years to the point where we can readily understand it. This point is perhaps best underscored by a famous quote by Isaac Newton, who is generally viewed as one of the greatest thinkers in the history of humankind:

> ***If I have seen a little further, it is by standing on the shoulders of Giants***

These few observations alone should be sufficient to disabuse you of the notion that you can go it alone. Although part of our country's early mythology glorified the rugged individualist, this image was always seen as a romantic ideal—nice to dream about but impossible to be.

If you have somehow been led to believe that working independently is the best way to approach your engineering studies, think again. Don't let such misconceptions stand in your way—as they undoubtedly will. Instead, take full advantage of the many resources and learning opportunities your campus offers you. The really smart student does.

At your college or university, there are two immediate resources available to help you with your academic work:

- **Your peers**
- **Your professors**

The value of making use of both your peers and your professors is best explained in the following excerpt from an excellent Harvard University study on the teaching/learning process [4]:

> Is there any common theme that faculty members can use to help students, and indeed that students can use to help themselves? The answer is a strong yes. All the specific findings point to, and illustrate, one main idea. It is that students who get the most out of college, who grow the most academically, and who are the happiest, organize their time to include interpersonal activities with faculty members, or with fellow students, built around substantive academic work.

I assume that you want to be one of those students "who get the most out of college, who grow the most academically, and who are the happiest." Now you know what you need to do: i.e., "organize [your] time to include interpersonal activities with faculty members, or with fellow students, built around substantive academic work." But you may not know how to bring about these results. In Chapter 4, we will provide you with specific strategies on making effective use of your professors, and in Chapter 5, we will address the important issue of making effective use of your peers.

REFLECTION

How do you feel about someone helping you with something? Are you a person who likes to help others? Do you like others to help you? If you get pleasure from helping others, why wouldn't you want to give others the same pleasure by letting them help you? Imagine that even after giving it your best effort, you just can't understand some material that was presented in a class. Who would you be most inclined to ask for help? Your professor? A classmate? No one?

3.8 Academic Success Skills Survey

Before you go on to the next chapter, complete the *Academic Success Skills Survey* at the end of this chapter. Note any items you checked as "Neutral," "Disagree," or "Strongly Disagree." Make a commitment to improve in those areas and as you study the subsequent chapters of this text give particular attention to the sections that address those areas in which you need improvement.

SUMMARY

This chapter provided an introduction to the *teaching/learning process*. You may not have previously given much thought to how this process works. Hopefully you now understand that the institution focuses primarily on the *teaching* part, while the *learning* part is left up to you.

We began by taking a look at the learning process. We defined learning and described its various component parts. You identified your preferred ways of receiving and processing new knowledge—information that can aid you in designing your learning process. We also encouraged you to continuously improve your learning process through "metacognition"—closely observing your learning process, feeding back what you observe, and making changes based on your observations.

We also provided you with a perspective on the importance of reinforcement in the learning process. Taking a systematic approach to your learning that involves repetition, review, and reinforcement will go a long way toward making you an "expert" learner.

Then we turned to the teaching part of the teaching/learning process. We discussed the various teaching styles used by your professors and the benefits to you of understanding how your professors teach.

Next we pointed out mistakes commonly made by students as they transition from high school into university-level engineering studies. Your success will depend to a great extent on your ability to avoid these mistakes.

We closed the chapter with a perspective on the concept of seeking help. I hope you were persuaded that "standing on the shoulders" of others is fundamental to the very concept of an education and that by recognizing it, you will take full advantage of all the resources available to you—particularly your peers and your professors.

In the next chapter, we will focus on what you can do to get the most out of the teaching part of the teaching/learning process. By practicing the approaches presented there, you will lay the strongest possible foundation on which to build your learning process.

REFERENCES

1. Anderson, L. W. and Krathwohl, D. R. (editors), *A Taxonomy for Learning, Teaching, and Assessing: A Revision of Bloom's Taxonomy of Educational Objectives, Complete Edition*, Longman, New York, NY, 2001.

2. Bransford, John, Brown, Ann L., and Cocking, Rodney R., *How People Learn: Brain, Mind, Experience, and School: Expanded Edition*, National Academies Press, 2000.

3. Felder, Richard M. and Silverman, Linda K., "Learning and Teaching Styles in Engineering Education," *Engineering Education*, v. 78(7), pp. 674-681, 1988.

4. Light, Richard J., *The Harvard Assessment Seminars: Second Report*, Harvard University, Cambridge, MA, 1992.

PROBLEMS

1. Review the 11 attributes presented in Chapter 1 that the ABET *Engineering Criteria 2000* mandates engineering graduates must have. Which primarily involve cognitive learning? Which are most likely to involve psychomotor learning? What psychomotor skills might you acquire? Which are most likely to involve affective learning? Which affective skills might you acquire?

2. Consider the 11 attributes presented in Chapter 1 that the ABET *Engineering Criteria 2000* mandates engineering graduates must have. In the table below indicate the extent to which each requires one of Bloom's higher level thinking skills—applying; analyzing; evaluating; creating. Fill in each box with a "1" – none or very little; "2" – moderate extent; "3" – significant extent.

	Applying	**Analyzing**	**Evaluating**	**Creating**
Apply knowledge of math, science, and engineering				
Design and conduct experiments				
Design a system, component, or process				
Function on multidisciplinary teams				
Identify, formulate, and solve engineering problems				
Understanding of professional/ ethical responsibility				
Communicate effectively				
Understand global and societal context				

Recognize need for lifelong learning				
Knowledge of contemporary issues				
Use techniques, skills, and modern engineering tools				

3. Go on line at: http://www.engr.ncsu.edu/learningstyles/ilsweb.html and complete the Index of Learning Styles Questionnaire developed by Barbara Soloman and Richard Felder at North Carolina State University. Write a two-page paper on why it is beneficial for you to understand your preferred learning styles. Include changes in your behaviors you plan to make based on this new information.

4. Review the list of characteristics of "expert" learners presented in the reflection exercise presented in Section 3.3. Divide the list into two categories: 1) those that describe you; and 2) those that don't describe you. Take the list of characteristics that don't describe you and rank the items in importance. Pick the most important item and develop a written plan that includes steps you can take to move this item to your other list.

5. How does the statement that "learning is a reinforcement process" match your past learning experiences? Can you think of a specific example where you learned something through repeated reinforcements? Can you think of a specific example of something you would have liked to learn but didn't because of inadequate reinforcements? Write down five things you can do to take advantage of this idea that you have not bccn doing on a regular basis.

6. Based on the five teaching styles presented in Section 3.4, determine how the way you prefer to learn compares to the way you are being taught. For each category, check the box that best describes the teaching style that is most prevalent in your math/science/engineering classes. Then check the box that describes the way you prefer to learn.

Type of Information Emphasized

Teaching style	Most prevalent in your classes	What you prefer
Concrete		
Abstract		

Mode of Presentation Stressed

Teaching style	Most prevalent in your classes	What you prefer
Visual		
Verbal		

Way Presentations Are Organized

Teaching style	Most prevalent in your classes	What you prefer
Deductive		
Inductive		

Mode of Student Participation

Teaching style	Most prevalent in your classes	What you prefer
Active		
Passive		

Type of Perspective Provided

Teaching style	Most prevalent in your classes	What you prefer
Sequential		
Global		

In what categories does the way you prefer to be taught match the way you are being taught? What are the implications for your learning? In what categories does the way you prefer to be taught differ from the way you are being taught? What are the implications for your learning?

7. Using your favorite Internet search engine, research one of the teaching approaches mentioned in Section 3.5:

   - Interactive lectures
   - Problem-based learning
   - Inquiry-guided learning
   - Just-in-time teaching

   Write a one-page paper describing the method.

8. Complete the *Academic Success Skills Survey* at the end of this chapter. Assign a point value to each question, based on the following point scale:

Strongly agree	+2
Agree	+1
Neutral	0
Disagree	-1
Strongly disagree	-2

   Compute your average score for the 16 statements in the survey. Then rate yourself as "outstanding," "good," "fair," or "poor" in practicing good academic success skills.

9. Pick six of the 16 areas in the *Academic Success Skills Survey* that you think are the most important for academic success. What is your average score for these?

10. From the six academic success skills you identified as most important in Problem 9, pick the two skills you feel you most need to improve. Develop a plan for improving in each area. Implement the plan.

ACADEMIC SUCCESS SKILLS SURVEY

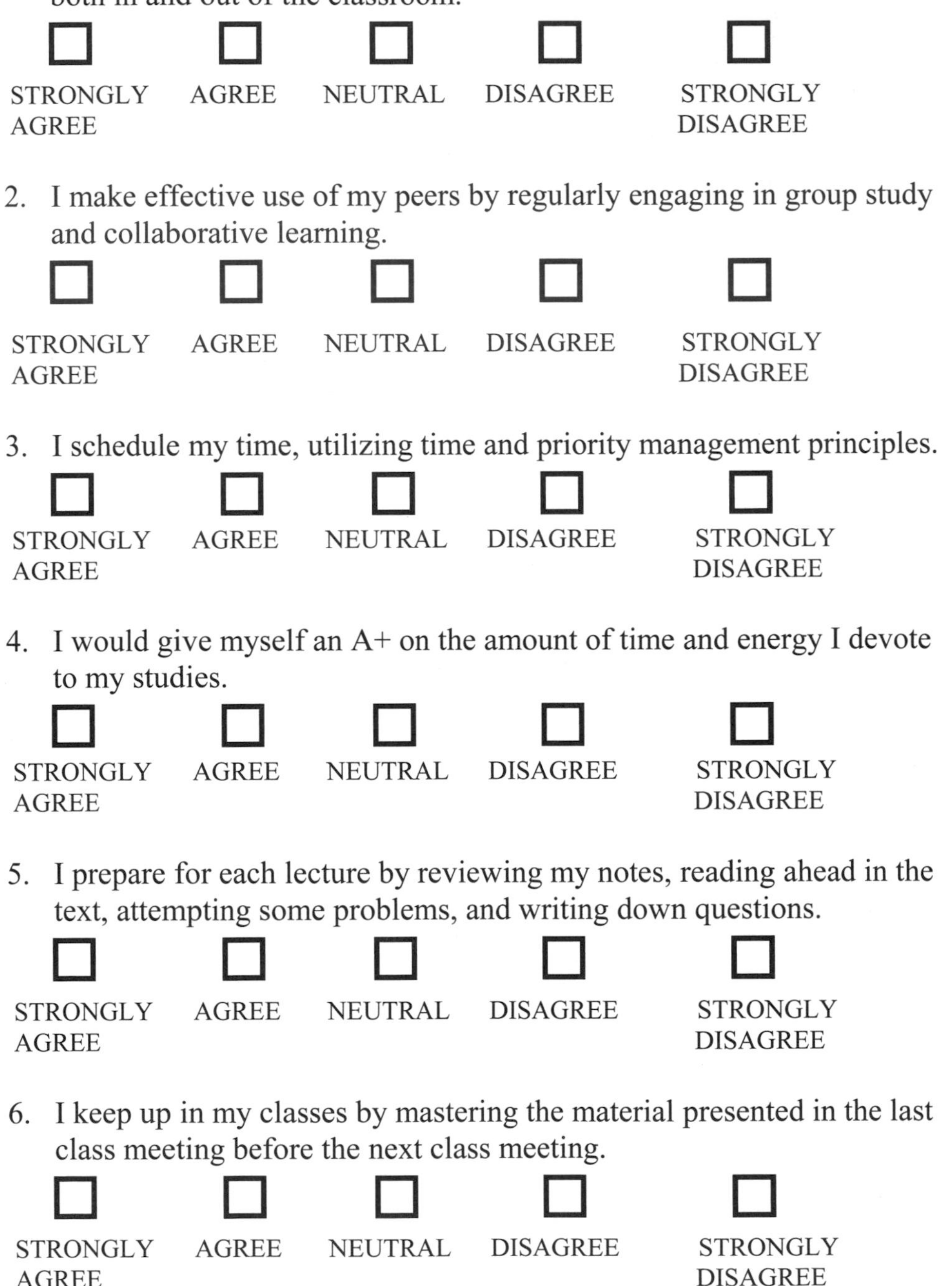

1. I interact regularly with my professors in positive, beneficial ways, both in and out of the classroom.

☐ STRONGLY AGREE ☐ AGREE ☐ NEUTRAL ☐ DISAGREE ☐ STRONGLY DISAGREE

2. I make effective use of my peers by regularly engaging in group study and collaborative learning.

☐ STRONGLY AGREE ☐ AGREE ☐ NEUTRAL ☐ DISAGREE ☐ STRONGLY DISAGREE

3. I schedule my time, utilizing time and priority management principles.

☐ STRONGLY AGREE ☐ AGREE ☐ NEUTRAL ☐ DISAGREE ☐ STRONGLY DISAGREE

4. I would give myself an A+ on the amount of time and energy I devote to my studies.

☐ STRONGLY AGREE ☐ AGREE ☐ NEUTRAL ☐ DISAGREE ☐ STRONGLY DISAGREE

5. I prepare for each lecture by reviewing my notes, reading ahead in the text, attempting some problems, and writing down questions.

☐ STRONGLY AGREE ☐ AGREE ☐ NEUTRAL ☐ DISAGREE ☐ STRONGLY DISAGREE

6. I keep up in my classes by mastering the material presented in the last class meeting before the next class meeting.

☐ STRONGLY AGREE ☐ AGREE ☐ NEUTRAL ☐ DISAGREE ☐ STRONGLY DISAGREE

7. I am aware of the importance of being immersed in the academic environment of the institution and spend as much time on campus as possible.

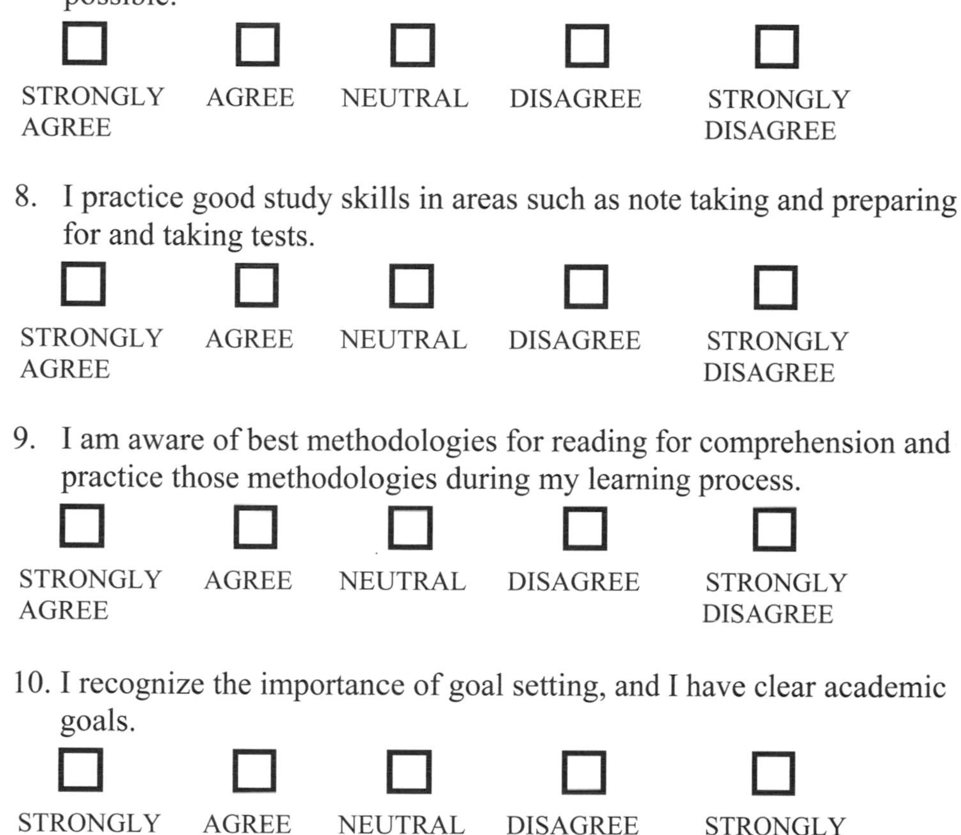

STRONGLY AGREE | AGREE | NEUTRAL | DISAGREE | STRONGLY DISAGREE

8. I practice good study skills in areas such as note taking and preparing for and taking tests.

STRONGLY AGREE | AGREE | NEUTRAL | DISAGREE | STRONGLY DISAGREE

9. I am aware of best methodologies for reading for comprehension and practice those methodologies during my learning process.

STRONGLY AGREE | AGREE | NEUTRAL | DISAGREE | STRONGLY DISAGREE

10. I recognize the importance of goal setting, and I have clear academic goals.

STRONGLY AGREE | AGREE | NEUTRAL | DISAGREE | STRONGLY DISAGREE

11. I am effectively managing the various aspects of my personal life, such as interactions with family and friends, personal finances, and outside workload.

STRONGLY AGREE | AGREE | NEUTRAL | DISAGREE | STRONGLY DISAGREE

12. I am highly motivated through a clear understanding of the rewards graduating in my chosen major will bring to my life.

STRONGLY AGREE | AGREE | NEUTRAL | DISAGREE | STRONGLY DISAGREE

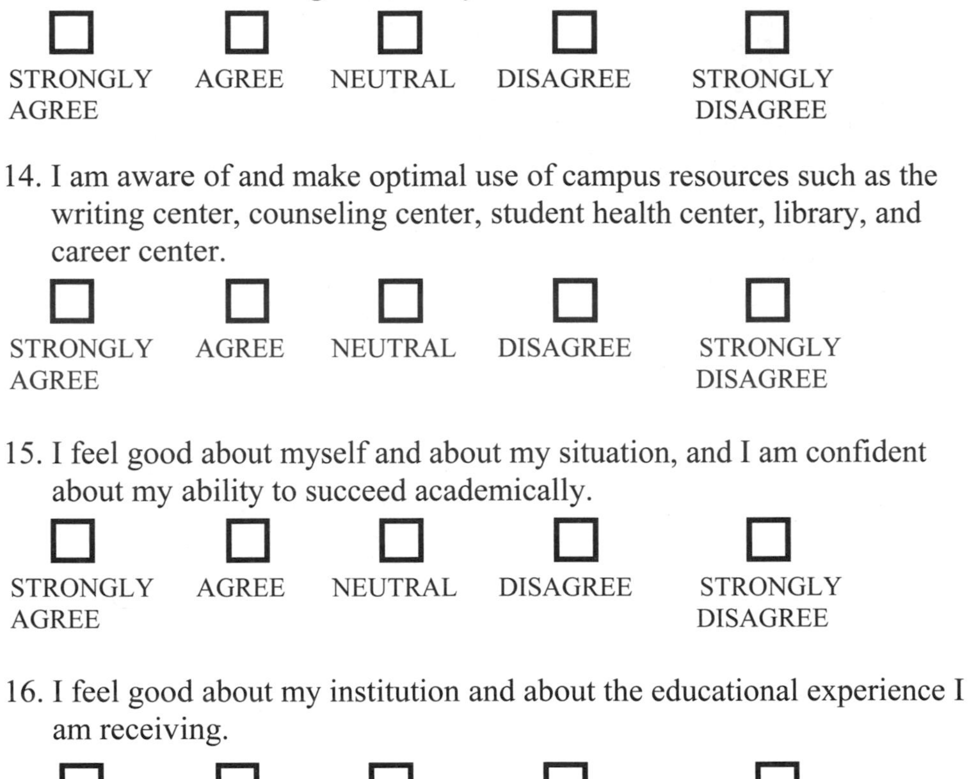

13. At my university, I know other students in my classes and feel part of an academic learning community.

☐	☐	☐	☐	☐
STRONGLY AGREE	AGREE	NEUTRAL	DISAGREE	STRONGLY DISAGREE

14. I am aware of and make optimal use of campus resources such as the writing center, counseling center, student health center, library, and career center.

☐	☐	☐	☐	☐
STRONGLY AGREE	AGREE	NEUTRAL	DISAGREE	STRONGLY DISAGREE

15. I feel good about myself and about my situation, and I am confident about my ability to succeed academically.

☐	☐	☐	☐	☐
STRONGLY AGREE	AGREE	NEUTRAL	DISAGREE	STRONGLY DISAGREE

16. I feel good about my institution and about the educational experience I am receiving.

☐	☐	☐	☐	☐
STRONGLY AGREE	AGREE	NEUTRAL	DISAGREE	STRONGLY DISAGREE

CHAPTER 4
Making the Most Out of How You Are Taught

You can observe a lot just by watching.

Yogi Berra

INTRODUCTION

In Chapter 3, we provided an overview of how your teaching is delivered. In this chapter we will focus on how to take full advantage of that teaching process. By doing so, you will ensure that you have a sound foundation on which to build your learning process. Specific approaches for designing your learning process are presented in the next chapter.

We begin this chapter by discussing early course preparation. We will emphasize that the "start" of a course is very important and that you need to be in the right courses, with the best available teachers, and have your textbook and other materials. We will also discuss the course syllabus as a potential source of important information.

We then present a number of important strategies and skills for taking full advantage of your lectures. One of the most powerful of these strategies is to prepare for lectures so that the lecture becomes a reinforcement of the material rather than an initial exposure. Other skills for getting the most out of your lectures including good listening skills, effective note-taking, and skill at asking good questions in class are covered as well.

Next, we present strategies for making effective use of your professors, another important resource both in and out of the classroom. Too often students either overlook or fail to understand the many benefits their professors can provide them. After describing these benefits, we will teach you how to build the kind of positive relationships with your professors that you will need in order to obtain these benefits.

We close the chapter with a section on "Utilizing Tutors and Other Academic Resources." Taking advantage of these resources will require initiative on your part, but the benefits can be worth it.

4.1 Early Course Preparation

The start of a course can be likened to the start of a race. When the starting gun is fired, you have to be off and running. Otherwise you will spend the whole race trying to catch up—something you are unlikely to be able to do. Your goal should be to be ready to "fire on all cylinders" from the get go. This means that you are in the right class, are mentally prepared, and have your textbook and other appropriate materials.

This all begins with the process of selecting your courses, ideally through an academic advisement session with your advisor. Being in the right class means that you have the appropriate background and prerequisites, and that the demands of the course, including its meeting time, fit into an overall manageable workload. When multiple sections of the same course are available to you, the selection of a specific section can include a comparative evaluation of the various instructors. Sources of information about instructors include other students, other professors, and in some cases published student opinion survey results.

Don't Be Lulled into Complacency

> Often a course starts slowly with only a small amount of material presented in the first class or even the first few classes. Don't use this as an excuse for getting off to a slow start. Remember what we said about this being the start of the race. Instead of being lulled into complacency, use the slow start to get on top of the course material. Later we will discuss the importance of mastering the material presented in each class before the next class comes. Make a resolve to adopt this approach from day one!

Mental preparation is not unlike getting "psyched up" for an important competition. As the start of the course approaches, check your mental frame of mind. Are you excited and focused? Are you clear that taking this class is important to you and that you want to be there? If not, remind yourself of what you learned in Chapter 2 about why you want to be an engineer and review how this course fits into your "roadmap" for accomplishing that goal.

USING THE COURSE SYLLABUS

Generally each of your professors will provide you with a course syllabus during the first week of the term. The syllabus can be a gold mine of valuable information. My advice is to study the syllabus thoroughly and keep it in a readily accessible place so you can revisit it frequently.

Syllabus Content

Course Information
Course title, course number, credit hours, prerequisites, classroom location, dates and times class meets

Instructor Information
Full name, title, office location, office phone number, e-mail address, office hours

Textbook(s)
Title, author, date (and edition), publisher, cost, extent to be used, other reference materials

Course Description/Objectives
Course description, instructional methods, content, goals, and objectives. Note: This item could range from as little as a repeat of the course description from the college catalog to a listing of detailed educational objectives, i.e., what students are expected to be able to do to demonstrate knowledge, skills, and attitudes learned in the course.

Course Calendar/Schedule
Daily (or weekly) schedule of topics, reading assignments, problem assignments, exam dates, due dates for assignments, special events (e.g., field trips, guest speakers, etc)

Course Policies
Attendance, lateness, class participation, missed exams or assignments, lab safety/health, emergency evacuation, academic dishonesty

Basis of Grading
Percentage of grade devoted to quizzes, final exam, homework, projects, essays and term papers, attendance, class participation

Available Support Services
Library references, learning center, computer resources

The syllabus should contain some or all of the above information. Since all of the items on this list are things you need to know, if any are missing from the syllabus, I would encourage you to fill in the missing items on your own.

Hopefully, this list has persuaded you of the importance of "mining" the syllabus for important information.

ACQUIRING TEXTBOOK AND OTHER MATERIALS

You should get your textbooks right away—perhaps soon after you register for a course. Don't wait until the term starts. You can benefit from scanning your textbooks and even studying the first few chapters during the break period preceding the start of the next term. Also, it is not uncommon for a campus bookstore to run out of books—and you can't afford to be without a book for the several weeks it might take for the additional books to arrive. If you do buy your books early, save your receipts and refrain from writing in the books so you can return them if necessary. If money is a problem, consider used books either from your bookstore or from Internet book dealers such as Amazon.com or eBay.com (although used books may be more difficult to return or exchange).

Make sure you have other materials you need as well. These would include a notebook for taking notes and other supplies and equipment such as a personal computer and/or a hand-held calculator.

4.2 PREPARING FOR LECTURES

Preparing for lectures is a powerful and effective strategy for success, and an excellent opportunity to reinforce your learning. It is unfortunate that so few students practice it—or even know how to do it—for it yields so many benefits. It's a little like "warming up" for a physical workout. Students who take time to prepare for their lectures go into the lecture with more interest, follow the lesson with more ease, and come away with more knowledge than those who walk in "cold."

All these benefits derive from its role in the "reinforcement" process of learning. If you study a lecture topic in advance, even briefly, the lecture becomes your first reinforcement, rather than your initial exposure to the subject. Thus, both your level of learning and interest are enhanced.

An excellent book titled *How to Study in College* by Walter Pauk and Ross J. Q. Owens [1] puts it this way:

> "Each lecture can be viewed as a piece of a puzzle. The more pieces you have in place, the more you know about the shape and size of the pieces that remain. The course syllabus, the notes from your last lecture, and related reading assignments can all function as these puzzle pieces as you prepare for a lecture."

While preparing adequately for your lectures does require effort on your part, it's not all that difficult or time-consuming. Prior to class—the night before or if feasible during the hour just before class begins, review your notes from the previous class, read over the next section in your text, try a few of the problems at the end of the chapter, and write down questions about things you're unsure of.

Try to do this for at least a week or two to see how such little effort can have such a big impact on what you get out of your lectures. I'm sure you'll be surprised by the results and subsequently make this part of your regular study routine.

REFLECTION

Think about going to a concert given by your favorite musical group. Which songs do you enjoy the most? Those that you have heard many, many times before? Or those you have never heard (e.g., from a new album)? Why do you think a person might enjoy and get more out of hearing things they've heard before? Do you believe these reasons carry over to the idea of preparing for your lectures?

4.3 DURING YOUR LECTURES

Once you have prepared for a lecture, there are several tactics that will help you get the most out of the lecture: sit near the front, concentrate on the material being presented, practice good listening skills, take thorough notes, and ask questions.

SIT NEAR THE FRONT

Studies show that students who sit near the front of the classroom perform better than those who sit in the back. Sitting near the front has several obvious but important advantages. You will hear better, see better, have fewer distractions, and be better positioned if you want to ask a question or otherwise get your professor's attention.

"BE HERE NOW"

Getting the most out of your lectures requires that you learn how to keep your attention focused—i.e., that you "be here now." This is not easy, as most students—indeed, most people—have short attention spans. From time to time, your mind will wander to other thoughts, thus tuning out the lecture and perhaps missing important points. When this happens, you need to immediately "slap yourself" mentally and return your attention to the lecture. Every time you do this, you will increasingly strengthen your ability to concentrate on the "here and now." (You'll find this ability extremely valuable not only in lectures but in many other situations, both as a student and later as a practicing engineer. Just one example is the need to "stay on task" when working in study groups, an important success strategy we will discuss in Chapter 5.)

LISTENING SKILLS

Good listening skills can be developed, but doing so is often overlooked. To a great extent, being a good listener means being an active listener. Listening is more than hearing. Listening is a conscious choice process. Once you know the difference between good listening habits and poor listening habits, you can choose one or the other. It's really up to you. Try them!

The following nine keys to effective listening should be enough to get you going. (Adapted from *How to Study in College* [2])

Poor Listener	Good Listener
Tunes out uninteresting and boring topics. Turns off quickly.	Works at finding value in all topics. Listens to discover new knowledge.
Tunes out if delivery is poor.	Judges value of the content rather than the delivery.
Listens for facts and details.	Listens for central themes. Uses them as anchor points for the entire lecture.
Brings little energy to the listening process.	Works hard at listening; remains alert.
Readily reacts with opposing views to new ideas. Starts listening to themselves when they hear something they don't agree with.	Focuses on understanding completely rather than coming up with opposing views.
Bothered by distractions.	Fights distractions; ignores bad habits of other students; knows how to concentrate.

Resists difficult material; prefers light recreational material.	Welcomes difficult material as exercise for the mind.
Interrupted by emotionally charged words or ideas.	Does not get hung up on emotionally charged words or ideas; listens with an open mind.
Daydreams and lets mind wander off with slow speakers or gaps in presentation.	Uses extra time to think more deeply about what the lecturer is saying; summarizes what has been covered

REFLECTION

For each of the nine items in the table above, decide which column best describes you as a listener during your lecture classes. For each item in which you describe yourself as a "poor listener," decide whether you would benefit from changing your habit to one of a "good listener." Make a commitment to the change and try it out for a week in your classes.

NOTE-TAKING

Another effective way to get the most out of your lectures is to take good notes. Your notes essentially create a record of what your professor feels is important, and that in itself is important for two reasons. First, many professors cover only certain portions of a textbook while, second, others present material that the text does not address. In either case, your notes will help you in knowing what to study for tests.

Tips or instructions on how to take good notes are difficult to give, for there is no one "correct" way to go about it. Your note-taking techniques will depend on a variety of factors, such as your own preferred style, the type of class, and the professor's teaching methods. But the following generalizations might be helpful to keep in mind:

(1) Note only important details: do not try to record everything the professor says.

(2) Include anything the professor writes on the board or conveys through visual aids (such as slides or overheads), for that usually signals "important details."

(3) Write down whatever you think you might encounter on the exam.

There are some additional "nuts and bolts" issues related to note-taking. One is whether to use a spiral notebook or a three-ring binder. Each offer advantages. With the three-ring binder, notes you take while reading the textbook, solutions to homework problems, or other reference materials can be easily inserted and integrated into your class notes. Another benefit is that you can spread your notes from a whole lecture out in front of you. The advantage of a spiral notebook is that you're not likely to lose or misplace anything that you have written in it.

CORNELL NOTE TAKING SYSTEM. Another important issue is how to best lay out each page of notes. In Chapter 5, we'll discuss ways for using your notes during your learning process. That process will be facilitated by leaving space for adding two things to your notes when you study and annotate them.

1) Questions that your notes answer (cue column)
2) A summary of the what is contained in each page of your notes (summary area)

A widely used scheme for accomplishing this is the **Cornell Note Taking Method** developed by Walter Pauk at Cornell University [3]. Using a standard 8-1/2" x 11" sheet of paper, create a 6" x 9" note-taking area on each sheet of by drawing a horizontal line 2" from the bottom edge of the page and a vertical line 2-1/2" from the left edge of the page. This will yield a sheet that looks like:

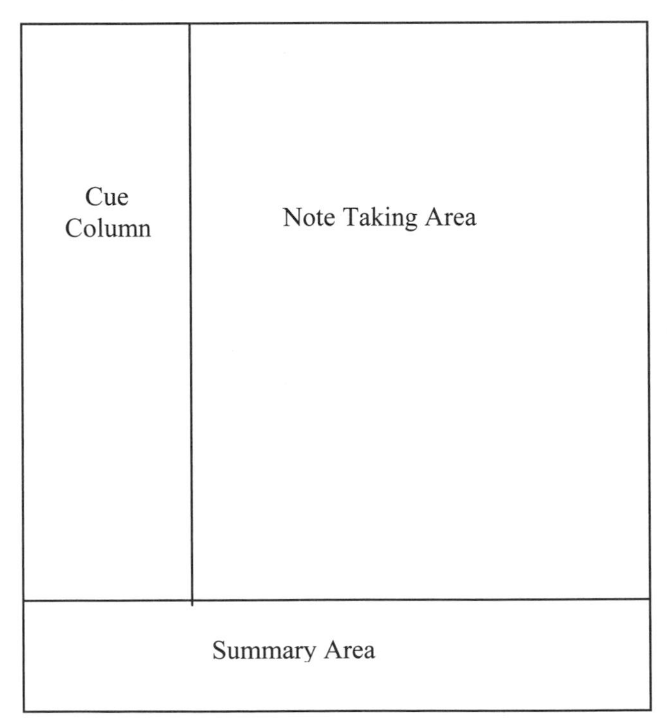

Take your notes as you normally would, but restrict them to the 6" x 9" note taking area. Leave the cue column and the summary area blank.

Later we will discuss how to use the blank areas during your learning process.

Whether you use the Cornell Note Taking format or just the "old fashioned" fill up one page and then go to the next page approach, remember that if you don't write something important down, it is unlikely you will be able to recall it later. Research in the cognitive processes of the brain has repeatedly shown that human memory is mostly short-lived. Unless an idea or information is consistently reinforced over a long period of time, it is quickly forgotten—usually in a matter of days. Your only alternative, then, is to record important information. That's why note-taking is an essential academic success strategy.

One last thought. More and more professors are turning to power point presentations for their lectures as an alternative to writing on a whiteboard or blackboard. Ideally, you will be given copies of the slides allowing you to concentrate mainly on listening to the lecture and jotting down a few important points rather than taking comprehensive notes. If you are not provided with copies of the power point slides, it will be virtually impossible for you both transcribe the slides and take notes on what the professor says, so do your best to capture the main ideas.

ASKING QUESTIONS IN CLASS

Another way to get the most out of your lectures is to ask questions in class. Don't be reluctant to ask questions. Many students are. When asked why, they give reasons like:

I was afraid I would look stupid
I didn't want to bring attention to myself.
I was too confused to know what to ask.
I didn't want to take time away from the instructor or other students.

Don't be one of these students. If you're confused, it's likely that many other students are as well. Just acknowledge it by raising your hand and when called on saying something like: "I'm confused about the last point you made." Take the view that the only dumb question is one that was never asked. And whose time do you think it is anyway? How about the view that it's your time? You deserve it!

Asking good questions is a skill that is not only useful in class, but as you'll see in later sections on learning, formulating questions can be an important tool in your learning process. Even further, improving your ability to ask good questions can benefit you in your career and in all aspects of your life.

The importance of this subject was well stated by the Greek philosopher Socrates more than 2,500 years ago:

The highest form of Human Excellence is to question oneself and others.

Because of the importance of questioning skills, we'll devote a bit more to it. There are many ways of categorizing questions. For our purpose, let's consider four categories:

Memory level questions

Convergent thinking questions

Divergent thinking questions

Evaluation thinking questions

Each of these categories is related to a different way we receive and process knowledge.

Memory level questions exhibit memory of previously learned material by recalling facts, terms, basic concepts, and answers. These questions usually begin with "Who," "What," "Where," or "When." Memory level questions often solicit a "Yes" or "No" or very short answer. Examples would be:

Who invented the electric light bulb?

What chapters will be covered on the mid-term?

Where do I go to sign up for academic advising?

When is the final exam?

Convergent thinking questions represent the analysis and integration of given or remembered information. These questions usually begin with "Why," "How," or "In what ways." Answers to these questions involve explaining, stating relationships, and comparing and contrasting. Examples would be:

Why do we use AC current rather than DC current in our homes?

How does a microwave oven work?

In what ways does the Linux operating system differ from Windows operating system?

Divergent thinking questions are those which represent putting information together in a different way by combining elements in a new pattern or proposing alternative solutions. These questions usually begin with "Imagine," "Suppose," "Predict," "If . . . ., then," "How might," "What are some of the possible consequences," "What could be changed to improve." These are often described as "open-ended" questions. Answers to these questions involve predicting, hypothesizing, inferring, reconstructing, or designing. Examples would be:

Suppose Einstein had never discovered his "Theory of Relativity." How would things be different now?

How might you go about improving your writing skills?

What could be changed to improve the gas mileage from SUVs?

What are some of the possible consequences of "global warming" on engineering job opportunities in the future?

Evaluation thinking questions are those which deal with matters of judgment, value, and choice. These questions usually begin with "Defend," "Judge," "Justify," "What do you think about," "What is your opinion about." Answers to these questions involve valuing, judging, defending, or justifying choices. Examples would be:

Would you justify why two years of calculus is required in the engineering curriculum?

What do you think about the new drop/add policy?

What is your opinion about the value of mandatory academic advising?

Why don't we generate more of our electricity using nuclear power plants?

I hope you'll use this model to improve your skill at formulating and asking questions. During class is one place you can put it into practice, but as we'll see in a later section, asking questions can be an important part of an effective learning process.

4.4 Making Effective Use of Your Professors

As discussed in the previous section, most of your professors are used to traditional teaching modes. Most are committed to a lecture style of teaching in which they convey knowledge to you in a one-way communication style. Most assign homework problems for you to do, collect and grade the problems, and so provide you with valuable feedback. Professors also determine your grade in the course—generally based on your scores on one or more tests and a final examination. This process of the professor lecturing, evaluating homework assignments, giving exams, and determining final course grades (which reflect the level of a student's mastery of the subject) is the standard mode of the teaching/learning process in engineering education.

IMPORTANT ROLES FOR YOUR PROFESSORS

But your professors can contribute much more than this to your overall education. The following is just a partial list of what professors can do for you:

- Give you the **benefit of the doubt** on a borderline grade
- Provide you with invaluable **one-on-one instruction**
- Give you **academic advising**, **career guidance**, and **personal advice**
- **Monitor your progress** and **hold you accountable** for your performance
- **Help you find a summer job** in industry and even **hire you** on their research grants
- **Serve as a valuable reference** when you apply for jobs, either while you are a student or after you graduate
- **Nominate you** for scholarships or academic awards

REFLECTION

Reflect on the above list of important roles for your professors. Are these things you would like your professors to do for you? Which ones would be particularly important to you? Would you like to have a close advisor or mentor? Would you like to have one-on-one instruction from an expert? Would you like to have a future reference for a job or scholarship? What would it take on your part to ensure that your professors will do these things for you?

VALUE OF ONE-ON-ONE INSTRUCTION. Of these, there is one in particular I'd like to expand on briefly, and that is one-on-one instruction—for this is probably the most valuable and beneficial role your professor can play outside of class.

One-on-one instruction is one of the best ways to learn, especially if the interaction is between an expert (i.e., teacher/professor) and novice (i.e., student). It is often referred to as the *Socratic method,* named after the great Greek philosopher Socrates, who used this method when he taught over 2,500 years ago. The primary advantage of the Socratic method is that the teacher can know immediately if the student understands the subject of their dialogue and, if necessary, adapt the lesson on the spot to ensure that the student truly learns it.

This teaching method would be ideal for engineering education—i.e., daily extended one-on-one meetings between just you and your professors—but realistically it is not possible. The most we can do is try to keep the teacher/student ratio as low as possible, while providing as many opportunities as possible for one-on-one instruction outside of the classroom or lecture hall.

One of these opportunities, and perhaps the best, is the weekly office hours that every professor is required to keep. In fact, the primary purpose of office hours is to give students the chance to work one-on-one with their instructors. If a student's schedule conflicts with his or her professor's office hours, as is often the case, most professors are willing to make appointments to meet with students at other times.

I urge you to use this opportunity regularly and frequently. As your education progresses, look for other opportunities to work one-on-one with your professors, such as offering to help them in their research projects or help out in their labs. Not only will such interactions enable you to learn more about engineering; they also will help you establish the kind of relationships with your professors you need to develop in order to derive the many other benefits that they can offer you.

TAKE RESPONSIBILITY FOR WINNING OVER YOUR PROFESSORS

To make effective use of your professors, you first must overcome any fear or intimidation of them you may feel. Being awed by your professors is a natural inclination since they are older and better educated, and often project a confident "know it all" attitude. As a result, you may think that your professors don't care about you—or even that they are somehow "against" you. But this isn't true. After all, most professors chose an academic career because they like teaching and enjoy working with students.

Remember, too, professors are human beings just like you, and that they have similar needs, fears, and insecurities as you. They may very much need to be liked, want you to think they are good teachers, need to impress you with their knowledge, or fear that they might make a mistake and reveal that they don't have a total command of their subject matter.

Once you get past any feelings of fear or awe, you need to realize that winning over your professors is your responsibility: you must take the initiative in establishing positive relationships with them.

How to Win Over Your Professors. The real question is how can you go about winning over your professors so that they want to help you. Perhaps the "bible" in winning people over is the classic book by Dale Carnegie: *How to Win Friends and Influence People* [4]. Written in 1936, this book has stood the test of time and is still a best seller. I recommend it to you as an excellent resource for you to improve your "people skills."

Dale Carnegie's "Six Ways to Make People Like You" lists helpful strategies that you can use to win over your professors:

Six Ways to Make People Like You

Rule 1	Become genuinely interested in other people.
Rule 2	Smile.
Rule 3	Remember that a person's name is to him or her the sweetest and most important sound in any language.
Rule 4	Be a good listener. Encourage others to talk about themselves.
Rule 5	Talk in terms of the other person's interest.
Rule 6	Make the other person feel important—and do it sincerely.

Dale Carnegie's book is filled with anecdotes. Most are dated, but their messages are timeless. The one I like the best is this story:

C. M. Knaphle, Jr., of Philadelphia, had tried for years to sell coal to a large chain-store organization. But the chain-store company continued to purchase its fuel from an out-of-town dealer and continued to haul it right past the door of Knaphle's office. Mr. Knaphle made a speech one night before one of my classes, pouring out his hot wrath upon chain stores, branding them a curse to the nation.

And still he wondered why he couldn't sell them.

I suggested that he try different tactics. To put it briefly this is what happened. We staged a debate between members of the course on "Resolved that the spread of the chain store is doing the country more harm than good."

Knaphle, at my suggestion, took the negative side; he agreed to defend the chain stores, and then went straight to an executive of the chain-store organization that he despised and said: "I am not here to try to sell coal. I have come to you for help because I can't think of anyone else who would be more capable of giving me the facts I want. I am anxious to win this debate; and I'll deeply appreciate whatever help you can give me."

Here is the rest of the story in Mr. Knaphle's own words:

I had asked this man for precisely one minute of his time. It was with that understanding that he consented to see me. After I had stated my case, he motioned me to a chair and talked to me for exactly one hour and forty-seven minutes. He called in another executive who had written a book on chain stores. He wrote to the National Chain Store Association and secured for me a copy of a debate on the subject. He feels that the chain store is rendering a real service to humanity. He is proud of what he is doing for hundreds of communities. His eyes fairly glowed as he talked; and I must confess that he opened my eyes to things I had never even dreamed of. He changed my whole mental attitude. As I was leaving, he walked with me to the door, put his arm around my shoulder, wished me well in my debate, and asked me to stop in and see him again and let him know how I made out. The last words he said to me were: "Please see me again later in the spring. I should like to place an order with you for coal."

To me that was almost a miracle. Here he was offering to buy coal without my even suggesting it. I had made more headway in two hours by becoming genuinely interested in him and his problems than I could have made in ten years by trying to get him interested in me and my coal.

I'm sure you get the point of this story; more importantly, I hope it has given you ideas as to how to approach your professors. This anecdote and Dale Carnegie's "Six Ways to Make People Like You" emphasize the importance of both showing interest in others and approaching them from their perspective.

REFLECTION

How do you relate to Mr. Knaphle's story? Have you had experiences where you came at people from your side of things and didn't get what you wanted from them? Have you ever tried approaching someone from their side of things? Who was it? Teacher? Parent or close relative? Friend? Co-worker? Boss at work? How could you apply the lessons of Mr. Knaphle's story to interacting effectively with your professors?

CHARACTERISTICS OF YOUR PROFESSORS YOU CAN COUNT ON. Just as Dale Carnegie knew Mr. Knaphle could win over the chain store executive by appealing to his interest in promoting chain stores, there are three characteristics of professors that you can almost always count on and therefore use to win them over:

(1) Professors think their areas of technical specialty are critically important and extremely interesting.

(2) Professors have elected an academic career over professional practice, and they believe they are outstanding teachers.

(3) Professors aren't called "professors" for nothing. They have big intellects and lots of knowledge, and love to convey what they know to others.

Your challenge as a student is to avoid doing anything that conflicts with these characteristics of professors; rather, think of ways to interact with your professors that reflect and reinforce these characteristics.

BEHAVIORS TO AVOID. We could make a long list of behaviors that conflict with professors' belief in the importance and interest of their technical specialties:

- Coming late to class
- Sleeping in class
- Talking in class
- Doing homework in class
- Using a cell phone in class to search the Internet, play video games, or do text messaging

- Leaving class early
- Failing to do the assigned homework

I'm sure you can add to this list.

The above behaviors also conflict with professors' belief that they are good teachers, as do other behaviors such as:

- Correcting professors' mistakes in an antagonistic tone
- Complaining that exams are too hard
- Complaining that grading is unfair
- Sending non-verbal messages to your professors that you dislike them personally

WINNING BEHAVIORS. Given what you now know, a good way to win over your professors is to send them messages that you find their subject both interesting and important, and that you value them as a teacher. You can start by practicing the opposite of the behaviors listed above. Be on time to class. Sit in the front. Pay attention. Ask questions. Apply yourself to the assigned homework.

But there is a much more direct way. Just tell them! In my experience, professors get far too few compliments. I'm not sure why students are so reluctant to tell their professors that they like the course, are interested in the subject, or appreciate the good job the professor is doing in teaching the class. I can assure you that doing so will go a long way toward winning over your professors.

And one final strategy for developing a positive relationship with your professors: show interest in them. In our Introduction to Engineering class, we give students an assignment to go visit one or more of their professors during their office hours and ask them questions about themselves. "Where did you go to college?" "How did you choose your technical specialty?" "How did you decide to become a teacher?" Students report very positive experiences from such interactions. Try it!

One additional bit of advice. Make the effort to know your professors' names. One of Dale Carnegies "Six Ways to Make People Like You" is:

Remember that a person's name is to him or her the sweetest and most important sound in any language.

I have frequently encountered students who couldn't tell me the name of their professors. Please don't be such a student.

Make Sure Your Professors Know Your Name. I'd like to raise one more issue related to names. Do your professors know your name? Probably not. Professors are busy people generally with too many students to learn and retain all of their names. But they do know some of their students' names and why shouldn't you be one of those? Set a goal of having your professors know your name and take responsibility for making it happen.

Reflection

What benefits will come to you if your professors know you by name that might not come to you if they don't? Do your professors in your key classes know your name? If you're not sure, how could you find out? If they don't know your name, what steps can you take to ensure that they do?

Understanding What Your Professors Do

You might benefit from understanding what university professors do. By doing so, you can be more sensitive to the demands placed on them and be more effective in building relationships with them.

University professors do more than teach classes. In fact, they are expected to perform in three primary categories (Note: the terms used to describe these categories may vary slightly from one institution to the next):

- Teaching
- Research
- Service

The **teaching category** includes not only classroom teaching, but also course and curriculum development, laboratory development, academic advising, and supervision of student projects or theses.

The **research category** includes creating and organizing new knowledge; disseminating and organizing new knowledge through publication of research papers, textbooks, software, and presentations at scholarly meetings; participation in professional societies and other activities that keep the faculty member up-to-date technically; and generating funds to support research.

The **service category** may include community involvement, participation in faculty governance through service on university committees, public service, consulting, and a variety of other activities.

Although most universities expect a faculty member to demonstrate accomplishments in all three categories, the relative importance given to each varies from one university to the next, depending on the characteristics of the institution. At one end are the so-called "research universities," which emphasize success in creating new knowledge, publishing the results, and obtaining funds to support these activities. Teaching loads at these institutions are relatively light—usually one or two courses a term.

At the other end of the spectrum are predominately undergraduate universities, which emphasize teaching. At these institutions, the teaching load is generally heavier—usually three to four courses a term. Some research or equivalent professional activity also is expected, but much less than at research institutions.

Communicating With Professors By E-Mail

Until recently, most communication between students and their professors occurred either in class (just before class, during class, or just after class) or during face-to-face meetings in the professor's office. E-mail has become a powerful tool for communication between students and their professors. Because of its potential for misuse and abuse, it is particularly important that you make judicious use of it. If your professor puts his or her e-mail address in the course syllabus, you can assume they will be receptive to your e-mail. If not, you should ask the professor whether they welcome e-mail communications from students.

A recent article in the *New York Times* about e-mailing professors [5] starts out:

> "One student skipped class and then sent the professor an e-mail message asking for copies of her teaching notes. Another did not like her grade, and wrote a petulant message to the professor. Another explained that she was late for a Monday class because she was recovering from drinking too much at a wild weekend party."

Hopefully, you wouldn't be inclined to send similar messages to your professors. Here are some guidelines that might help you with the protocols for e-mailing your professors.

- **Write from your college or university e-mail account** – Shows that your e-mail is legitimate and not spam.

- **Include the course number in your subject line** – Relieves your professor of the chore of figuring out what class you're in.
- **Choose an appropriate greeting** – "Hi/Hello Professor _____" is always appropriate. Don't use "Hey _______." Avoid first names unless you have been specifically invited to use them. Avoid "Dr" unless you are sure the professor has a doctoral degree.
- **Things to avoid**

  Rote apologies (better to relate any sad or serious circumstances in person).

  Unexpected attachments.

  Criticism of the professor or other students.

  Requests for information you can find from other sources.

  E-mail abbreviations and jargon you might use with a friend.

  Anything you would not say in a face-to-face meeting with your professor.

  Making unreasonable demands on your professor's time.
- **Things to do**

  Ask politely.

  Proofread what you've written.

  Sign with your full name, course number, and meeting time.
- **When you get a reply, say thanks** – Just hit "Reply" and say "Thank you" or a little bit more if that's appropriate. Try to avoid sending a second message that requires the professor to respond again.

Although wide-scale adoption of e-mail started as early as 1993, it is still developing as a tool for communication between students and their professors, and both groups are still learning how to best use e-mail to mutual benefit in the teaching/learning process. You can make a positive contribution to this process by following the guidelines presented above.

4.5 Utilizing Tutors and Other Academic Resources

Your university or college offers a number of academic services to support your education. Examples of these are tutoring, recitation sections of large lectures, and other academic services. These services are generally free to you, since you "pay" for them in advance through your tuition and student fees. However, receiving the benefits of these academic services requires that you take the initiative. They will not seek you out. Part of good academic gamesmanship, then, is for you to find out about the resources available to you and make optimal use of them.

TUTORING

Tutors are another excellent source of the type of one-on-one instruction discussed previously. Some students are reluctant to utilize tutors, equating the need for tutoring with an admission they are doing poorly or need help.

After what we've said about the myth of "succeeding on your own," you should realize how unfounded and counterproductive such reluctance is. If, however, you find yourself in this bind—in need of help but resistant to seek tutoring—try looking at tutoring in a more positive light: as an opportunity for you to have a dialogue with an expert on a subject you want to learn.

Your university may provide tutoring services through a variety of sources. Tutoring may be available through a campus-wide learning assistance center. Your mathematics department may run a math lab, or members of your engineering honor society, Tau Beta Pi, may do voluntary tutoring as a service to the engineering college.

If free tutoring is not available, you might find listings for tutors available for hire at your career center. Or you could just ask an upper-class student to help you. Lots of students like to "show off" their knowledge.

RECITATIONS/PROBLEM SOLVING SESSIONS

A common teaching mode is large lectures (100 to 300 students or more) accompanied by small recitations. Recitations are generally taught by graduate students serving as teaching assistants. The purpose of the recitation is to amplify and reinforce the main concepts from the lecture and to focus on working problems. Whereas there is little opportunity for you to ask questions in large lectures, you will find that there is more than ample opportunity to ask questions in recitations. To get the most out of recitations, it is important that you have studied the material presented in the lecture and attempted the assigned problems.

OTHER IMPORTANT ACADEMIC RESOURCES

Among the other academic resources available to you, here are some that you should certainly look into and use.

- The **academic resource center** can provide you with tutoring and help you improve your reading, writing, and study skills.
- Your **university library** not only maintains important books, periodicals, on-line material, and other references to support your engineering education, it also holds workshops and seminars on how to access all these sources of information. Reference librarians are also available through e-mail, telephone, or in person to assist you in identifying and retrieving information for research, study, or personal use.
- **Student computer labs** provide access to computer hardware and applications software, access to the Internet, resource materials, and training.
- **Academic advising** provides you help in reviewing your overall academic plan and your progress in following it and guidance in selecting courses for the next term. More detail on this important academic resource can be found in Chapter 8.
- The **university/college catalog** provides you with important academic information including rules and regulations, college and department information, curriculum, and course descriptions.
- Your **registrar's office** can help you with various academic procedures, including changing majors, dropping and adding classes, making grade changes, and transferring course credits from other institutions.

To find out what your campus offers and where these services can be found, check your university or college web page, catalog, and your department or school student handbook. Freshman orientation programs are also helpful for learning about student services. And don't forget one of your best resources: other students.

SUMMARY

In this chapter, we presented strategies and skills for making the teaching process work for you. We began by likening the start of a course to the start of a race, and pointed out that you need to be ready to go as soon as the "starting gun is fired." Strategies for ensuring that were presented.

Then we discussed a number of important strategies and skills for taking full advantage of your lectures. One of the most powerful of these strategies is to prepare for lectures so that the lecture becomes a reinforcement of the material rather than an initial exposure. Other skills for getting the most out of your lectures such as good listening skills, effective note-taking, and skill at asking good questions in class were covered as well.

We next explored the contributions that faculty can make to the quality of your education, both in and out of the classroom. We explained that deriving these benefits is your responsibility to pursue, and presented a variety of strategies and approaches for you to take in order to establish the kind of positive relationships with your professors you will need to receive these important "extras."

In addition to the support that your professors can offer you, we listed many other campus academic resources that can provide you valuable support. But once again, you must assume responsibility for seeking them out and taking advantage of them.

In the next chapter, we focus on how to make the learning process work for you. If you have implemented the skills and strategies presented in this chapter, you should have a strong foundation on which to build that process.

REFERENCES

1. Pauk, Walter and Owens, Ross J.Q., *How to Study in College, Eighth Edition*, p. 198, Houghton Mifflin Company, New York, NY, 2005.
2. Pauk, Walter, op cit, p.194.
3. Pauk, Walter, op cit, pp. 207-212.
4. Carnegie, Dale, *How to Win Friends and Influence People*, pp. 65-66, Simon and Schuster, New York, NY, 1936.
5. Glater, Jonathan D., "Emailing Professors: Why It's All About Me," *New York Times*, February 21, 2006

PROBLEMS

1. In the section on early course preparation, an analogy was made between the start of a course and the start of a race. Think about the idea that the start of things can be very important. Make a list of examples of things in which the start is extremely important. For each item on the list, consider whether a poor start can be overcome and what it would take to do so.

2. How do you decide which section to take in courses in which multiple sections are offered? Do you work proactively to get in the section having the best teacher? What do you do? What are some strategies you could adopt that you haven't been using?

3. Review the course syllabus in at least one of your key courses. Compare the information there to the list of items presented in Section 4.1 on "Using the Course Syllabus." Are most or all of the items included? If not, which ones are missing? Take on the task of filling in the missing items.

4. Make a commitment to prepare for your lectures using the approach discussed in Section 4.2 for a period of two weeks. At the end of the two weeks, write a two-page paper discussing the benefits you received (or didn't receive) from doing this.

5. Review the characteristics of "poor" listeners and "good" listeners presented in section on "Listening Skills" in Section 4.3. Pick two or three items from the "poor" listener list that describe you, if only occasionally. Make a commitment to try to change the habit to that described in the "good" listener column. Try out the new habit for two weeks and reflect on how it worked.

6. Try out the Cornell Note Taking System presented in Section 4.3. Check if your bookstore sells pre-made forms for this purpose. If not, make up your own forms. Preview the information on studying and annotating your notes in Section 5.2 of the next chapter. Make a commitment to follow the process of annotating your notes including developing questions in the "cue column" and summarizing each page in the "summary area." Try answering the questions in the cue column out loud over a period of two weeks and reflect on how it worked.

7. Practice your questioning skills by making up three additional questions that would be appropriate for inclusion at the end of this chapter. E-mail them to your Introduction to Engineering course instructor and to me at: rlandis@calstatela.edu.

8. For one of your key classes, make up two questions that fit each of the four categories of questions presented in Section 4.3 (memory level; convergent thinking; divergent thinking; evaluation thinking). Take them with you to the next lecture and see if you have an opportunity to ask one or more of them.

9. Pick one of the important academic success skills below and conduct a search on that skill using your favorite Internet search engine.

Note taking skills
Listening skills
Questioning skills

Gather information from a number of sites and write a 250 word paper on what you learned about the skill.

10. Make a list of behaviors that would send signals to your professors that you don't think their technical specialty is either interesting or important. Do you engage in any of these behaviors? Which ones?

11. Explain how the skills you develop in learning how to make effective use of your professors will have a direct carryover in the engineering work world.

12. Do you think that grading is objective or subjective? Ask two of your professors how they go about making up their final grades. Ask them what factors they consider in deciding borderline grades (e.g., *A/B, B/C, C/D*). Are these factors objective or subjective? Write a one-page "opinion report" on what you learned.

13. Do you believe that you will bother your professor if you go to his/her office to ask questions? If so, why do you believe this?

14. Go see one of your professors during his or her office hours. Ask one or more of the following questions:

    a. Why did you choose teaching as a career rather than professional practice? Would you recommend an academic career to others? Why or why not?

    b. Would you advise me to continue my engineering education past the B.S. degree? What are the advantages of getting an M.S. degree? A Ph.D. degree?

    c. I understand that your technical specialty is in the field of ________________. How did you get interested in that field? Do you think it would be a good field for me to consider?

    d. What do you think are the most important factors in an engineering student's academic success?

15. Make up five additional questions like the ones in Problem 14 that you could ask one of your professors. Pick the three you like the best and ask them of one of your other professors.

16. Look into the availability of free tutoring services on your campus. Are there tutors to help you with your mathematics classes? Are there tutors to help you write a term paper?

17. Pick one of the following offices on your campus. Stop by and seek information about the academic services offered there. (Note: The specific names may vary from campus to campus.)

    Academic Resource Center
    Library Reference Desk
    Open Access Computer Laboratory
    Registrar's Office
    Academic Advising Center

    Prepare a two-minute presentation on what you learned for your next Introduction to Engineering class meeting.

CHAPTER 5
Making the Learning Process Work For You

To improve is to change.
To be perfect is to change often.

Winston Churchill

INTRODUCTION

In Chapter 3, we provided overviews of how your teaching is delivered and how your learning takes place. In Chapter 4, we presented approaches for taking full advantage of the teaching process. The focus of this chapter is on how to design your learning process.

We begin the chapter by discussing two important skills for learning—reading for comprehension; and analytical problem solving. Your effectiveness as a learner will depend to a great extent on how well you do in developing your skills in these two areas.

Next we provide you powerful principles and approaches for organizing your learning process. The importance of keeping up in your classes is discussed and steps for mastering the material presented in each class are presented. Mastering the material presented in each class before the next class comes requires a strong commitment to both time management and priority management, so these important topics are discussed in detail.

We then describe approaches for preparing for and taking tests. Since the primary way you will be called on to demonstrate your learning will be through performance on tests, it is imperative that you become good at doing so.

The chapter concludes with a detailed discussion on making effective use of one of the most important resources available to you—your fellow students. Working collaboratively with your peers, particularly in informal study groups, sharing information with them, and developing habits of mutual support, will be critical factors in your academic success and the quality of education you receive.

5.1 Skills for Learning

We begin this chapter by discussing two important skills for learning: 1) reading for comprehension; and 2) analytical problem solving.

Reading for Comprehension

Much of your learning will involve comprehending information presented in written materials. And much of this material will be highly technical in nature. In fact, a typical four-year engineering curriculum will include about three years (96 semester credit hours) of technical courses (math, physics, chemistry, engineering, computing) and one year (32 semester credit hours) of non-technical (humanities, social science, communication skills) courses.

Although the methodology you learn in this section can be applied to both your technical and non-technical courses, it is particularly important for your technical courses. One important difference is "speed." You may equate having "good" reading skills with being a fast reader and even consider taking a "speed reading" course. Speed reading may be helpful in reading a novel for pleasure or for reading the morning newspaper, but will not be of much use in your technical courses. Mastering mathematics, science, and engineering content is generally a slow, repetitive process that requires a high level of active participation on your part.

There are a number of methodologies for reading for comprehension. All involve developing your skills in three areas:

- What you do ***before*** you read
- What you do ***while*** you read
- What you do ***after*** you read

Before You Read. The amount you learn from a reading task can be greatly enhanced by taking a few minutes to do three things before you start reading:

1. **Purpose** – Establish a purpose for your reading. The purpose might be entertainment or pleasure. Or it might be to find out one single piece of information. But in technical courses more often than not it is to comprehend principles and concepts that will enable you to solve problems at the end of a section of the book or the end of a chapter.
2. **Survey** – Decide on the specific scope/size of the reading (1 page; 1 section; 1 chapter). Survey/skim/preview that page or section or chapter. Only devote a few minutes to this process. Look at headings and subheadings. Inspect drawings, diagrams, charts, tables, figures, and photographs. Read the introductory section or paragraph and the summary section or paragraph.
3. **Question** – Write down a list of questions that you want to answer from the reading. A useful technique is to turn section headings and subheadings into questions. For example, some questions you might have written down for this section are: What do I need to learn to do before I read? What are the benefits of doing these things? What are some of the things that might keep me from doing them?

While You Read. The following is a list of suggestions for improving your comprehension when reading technical material.

- Never sit down to study without a paper and pencil at hand. You'll need them for sketching graphs, checking derivations, summarizing ideas. This kind of active reading is very important.

- Focus on concepts, not exercises or problems. The goal of most technical coursework is to guide you in understanding concepts that can be applied to a variety of problems. Rather than focus on how one particular problem is solved, focus on first understanding the general concepts thoroughly. Pay close attention to the mathematical formulas. Work carefully through each derivation. Take time to absorb graphs and figures. One of the biggest mistakes students make is to skim over the material in order to get on with the homework problems. Don't bypass the learning process in your rush to get your assignment done.

- Don't try to read too fast. In a half hour, you might read 20-60 pages in a novel. You might spend the same half hour on a few lines of technical material. Mathematics says a lot with a little. Become an active participant in your learning process. At every stage, decide whether the idea presented was clear. Ask questions. Why is it true? Do I really believe it? Do I really understand it? Could I explain it to someone else? Do I have a better way or method of explaining it?

- Write down anything that you don't understand. Where possible, put it in the form of a question. Seek to answer such questions by re-reading the text or using alternate sources such as other textbooks or the Internet. Pose the questions to other students or to your instructor during his or her office hours. As you read, there may be questions that are piqued by your curiosity but are not answered in the reading material. Take on the challenge of finding the answers to some of these questions.

- Periodically stop reading and **recite** what you have read. Using your own words, repeat to yourself (ideally aloud) what you have read. This is perhaps the most important step in the reading process. If the material is difficult, you may want to recite after reading only a few paragraphs. The purpose of reciting is for the knowledge to get implanted in your brain. Do not look at the book as you recite. If you can't remember what you just read, re-read the material.

AFTER YOU READ. Once you have read your text, most of the learning is still ahead of you. The following are three important tasks to perform after your reading—recite and reread; review; and solve problems.

- **Recite and Reread**. – Recall that during the process of preparing for your reading, you formulated questions you would like to have answered by the reading. Recite answers to those questions. If you need to, reread sections of the text. Again, it is preferable to recite aloud. Even better, recite to others. One of the best ways to learn anything is to teach it to someone else. Form a study group or meet with a study partner and practice teaching each other what you have learned from the reading (more detail on group study will be presented in Section 5.4). You can even tell friends and family members what you are learning. Talking about what you have learned is a powerful way to reinforce it.

- **Review** – Recall from Chapter 3 that learning is a reinforcement process. Only through repeated exposure to knowledge can we move it from our short-term memory to our long-term memory. Plan to do your first complete review within one day. Review the important points in the text and recite some of the main points again. Do it again in a week and then when you prepare for a quiz or exam and again when you prepare for the final examination.

• **Solve Problems** - Once you have read your text for comprehension, it's time to work problems. Being able to solve problems with speed and accuracy is to a great extent what you will be judged on in your math/science/engineering coursework. This requires both a systematic problem solving approach and lots and lots of practice. You can't work too many problems. First do any assigned problems. But don't stop there. If possible, work all the problems in the book. If you still have time, work them again.

Most of the problems you will encounter in math/science/engineering coursework can be described as "analysis" problems. Because of the importance of this type of problem in your education, the next section discusses a methodology for analytical problem solving.

EXERCISE

Apply the reading methods you have just learned to the following passage:

Shortcut For Adding Consecutive Numbers Starting with 1

The following theorem was published in Levi Ben Gershon's 13th Century manuscript Maaseh Hoshev (The Art of Calculation).

"When you add consecutive numbers starting with 1, and the number of numbers you add is odd, the result is equal to the product of the middle number among them times the last number."

In the language of modern day mathematics, this is written as:

$$\sum_{i=1}^{2k+1} i = (k+1)(2k+1)$$

Before reading, did you establish a purpose for the reading? Did you first skim the passage? Did you write down any questions you wanted answered? During reading, did you follow any of the suggestions made in the previous section? Most importantly did you recite what you learned? Can you apply the theorem? For example, how quickly can you add all consecutive numbers from 1 to 99? What level of understanding did you achieve? What does the symbol 'k' represent? Can you find a better way to represent the theorem mathematically? Were any unanswered questions piqued by your curiosity? Do you want to know more about Levi Ben Gershon? Did learning about this theorem pique your interest in other similar theorems?

PROBLEM SOLVING

Engineers are problem solvers. Much of your engineering education, and indeed your engineering career, will be devoted to improving your ability to think both logically and creatively to solve problems. There are many types of problems. Some examples would be:

Type of Problem	Example
Mathematics problems	Determine the probability that two students in a class of 30 have the same birthday.
Science problems	Find a theory that explains why a dimpled golf ball travels further than a smooth one.
Engineering analysis problems	Find the stress in a beam, the temperature of an electronic component, or the voltage at a node in a circuit.
Engineering design problems	Design a better mousetrap, a car alarm that goes off if the driver falls asleep, a better device for waiters and waitresses to carry plates of food from the kitchen to the table.
Personal problems	Health problems, financial problems, relationship problems.
Societal problems	Global warming, immigration, health care, world hunger.

GENERAL PROBLEM SOLVING METHODOLOGY. Because there are so many types of problems, there are many different problem solving methodologies. For example, the engineering design process was described in Chapter 2. You may be familiar with the "scientific method" in which a hypothesis is developed to explain some observed physical phenomena and then tested through experiment.

A good general approach for solving problems involves the following steps:

- Figure out where you are (problem definition)
- Figure out where you want to be (need or opportunity)
- Determine what resources are available
- Identify any constraints

- Develop alternate solutions that get you from where you are to where you want to be while staying within available resources and not violating any of the constraints
- Choose the best solution
- Implement the solution

This systematic approach can be applied to such varied problems as buying a new TV, dealing with your car breaking down in the middle of no-where, finding a job, running for president of your engineering student organization, starting your own company, stopping the re-zoning of an area of your neighborhood, or designing a better mousetrap. All of these examples can be described as open-ended problems, meaning that they have no single right answer or solution.

ANALYSIS PROBLEMS. Much of your engineering education, however, particularly in the first several years, will involve one specific type of problem—analysis problems. Generally such problems have a single "right" answer. *Analysis* typically involves translating a physical problem into a mathematical model and solving the resulting equations for the answer. The problem statement will be provided to you by your instructor either as a handout or from your textbook. The principles you need to solve the problem will generally be contained in your text material, although you may need to draw on knowledge from prerequisite courses.

Your success in engineering study will depend to a great extent on your ability to solve such problems accurately and often under time pressure. Problem solving is not a science but rather an art. It involves learning, thinking, logic, creativity, strategies, flexibility, intuition, trial and error. Even so, becoming a proficient analytical problem solver can best be accomplished if you adopt, practice, and become proficient at the following four-step systematic approach adapted from Polya [1]. This approach is not an algorithm (i.e., a series of steps that if applied correctly are guaranteed to lead to a solution) but rather a heuristic (a general set of guidelines for approaching problem solving that do not guarantee a solution).

Step 1. Understand the problem. Read the problem carefully. Identify the question you are being asked to answer. Identify the unknown(s) and assign each unknown a symbol. List all known information. Draw a figure, picture, or diagram that describes the problem and label it with the information you have extracted from the problem statement.

Step 2. Devise a plan. The goal of this step is to find a strategy that works. Because there are many possible approaches, this is perhaps the most difficult step in the problem solving process. Think about possible relationships between the known information and the question you need to answer. Depending on the nature of the problem, the following is a list of problem solving strategies to try:

Solve a simpler problem
Make an orderly list
Look for a pattern in the problem
Draw a diagram
Use a model
Use a formula
Work backward
Make a table
Guess and check
Eliminate possibilities
Solve an equation
Use direct reasoning
Consider special cases
Think of a similar problem
Solve an equivalent problem

Step 3. Carry out the plan. Implementation of the plan depends on the nature of the problem and the problem solving strategy chosen. In all cases, work carefully and check each step as you proceed. For many analysis problems by this point, you should have reduced the problem to essentially a purely mathematical one. Work through each step in any mathematical manipulations or derivations. Complete any required calculations using an electronic calculator or computer. Take particular care to ensure correct handling of units, a frequent source of errors in engineering problem solving.

Step 4. Look Back. Examine the solution obtained. Make sure your solution is reasonable. Recheck your calculations and review your reasoning. If possible, check your answer to verify that it agrees with the information given.

5.2 ORGANIZING YOUR LEARNING PROCESS

The following sections provide you with perspectives on the most important aspects of organizing your learning process.

"TAKE IT AS IT COMES"

I use the expression "take it as it comes," an axiom you've undoubtedly heard before, to emphasize what I consider to be the key to success in mathematics, science, and engineering courses. Stated more explicitly:

> **Don't allow the next class session in a course to come without having mastered the material presented in the previous class session.**

Because, to me, this is the single most powerful academic success strategy of all, ***if you are willing to put only one new behavior into practice, this is the one to choose!***

Have you ever wondered why a typical course is scheduled to meet only one, two, or three times a week rather than all five days? And why the total weekly hours of class meetings are so limited? After all, if you met nine hours a day for five straight days, you could theoretically complete an entire course in one week—and cut down the time to complete your undergraduate degree from four years to less than one year!

The answer is obvious. The teaching part of the teaching/learning process could be compressed into one week, but certainly not the learning part. You can only absorb a certain amount of material at one time, and only when that material is mastered can you go on to new material. Thus, your institution has designed a sound educational system in which professors sequentially cover small amounts of material for you to master. However, unless you do your part, you can easily turn that sound educational plan into an unsound one.

REFLECTION

Reflect on your approach to your classes. Do you operate on the principle that you master the material presented in each class before the next class comes? Or do you put off studying until a test is announced and then cram for the test? If you are in the second category, are you willing to change? What steps do you need to take to make that change?

PROCRASTINATION

Most students make the mistake of studying from test to test rather than from class to class. In doing so, they fall victim to a student's greatest enemy—procrastination. Procrastination is an attitude that says,

"Do it later!" "Doing it later" rarely works in any courses, but especially not in math, science, and engineering courses, in which each new concept builds on the previous ones.

If you are a procrastinator, for whatever reason, you are ignoring the sequential nature of engineering study, as well as your own inability to absorb complex information all at once. So you can't realistically expect to succeed if you delay your studying until a test is imminent. That's why I tell you to "take it as it comes."

A Common Trap

One trap you can fall into is a false sense of security because the teacher presents the material so clearly that you feel you understand it completely and therefore do not need to study it. But when you attend a lecture that is presented clearly, it only proves that the teacher understands the material.

What is necessary is for you to understand it—for you to be able to give the lecture. In fact, that should be your goal in every class: get to the point where you could give the lecture.

Often I am invited to be a guest speaker at Introduction to Engineering courses that use this text. Because it's such an important topic, I always make it a point to address the subject of *procrastination.* Typically, I'll ask the class: How many of you would say–"I am a procrastinator?" To this day, I'm surprised that not only do almost all the hands go up, but they go up enthusiastically (almost proudly) and are held high. It seems as though there's pride in being a member of a club—the "procrastinator club."

You may think of procrastination as doing nothing. But we're always doing something. Ultimately procrastination involves choosing to put off something we know we should be doing and instead doing something we know we shouldn't be doing. And why would we do such a thing? Why would we delay an action we know we should be doing? There are lots of reasons:

Fear of failure – Task is perceived as too difficult. "If we don't try it, then we haven't failed."

Fear of success – Accomplishing task might be resented by others. Or success might bring responsibilities and choices that we view as threats or burdens.

Low tolerance for unpleasant tasks – Task is viewed as not being pleasant. Doing the task may bring some discomfort.

Disorganized – Prefer to spend time worrying about not doing than doing. Unwilling to set priorities, develop a schedule, and stick to the schedule.

There are many strategies for dealing with procrastination. If you find you're putting something off because you think of it as unpleasant, a good approach is the 10-minute rule. Acknowledge, "I don't feel like doing that," but make a deal with yourself and do it for 10 minutes anyway. After being involved in the activity for 10 minutes, then decide whether to continue. Once you're involved, it's easier to stay with a task. And if you're overwhelmed with the difficulty of a task, use the "Swiss Cheese Method." Poke holes in the big project by finding short tasks to do that will contribute to completion of the larger project.

REFLECTION

Recall one or two recent examples where you put off an activity or task you should have done. Describe each activity you put off. What were you feeling when you chose to put off each activity? What were you thinking when you first chose to delay? Can you identify a discomfort that you were avoiding by putting off the activity?

MASTERING THE MATERIAL

As previously discussed, you will learn better if you "take it as it comes," mastering the material presented in each class session before the next session comes. In fact, research on learning indicates that the sooner studying takes place after the initial exposure to the material, the more fortified the learning will be. Having a study session right after class would be ideal, but if that's not possible, doing it the same evening would be better than the next day. Since your goal is to master the material, start by studying and annotating your notes, reading (or rereading) the relevant portions in your text, and working problems—as many as you can.

LEARNING FROM YOUR LECTURE NOTES. Recall in the section on the **Cornell Note Taking Method**, we recommended that you structure your notes to leave two areas blank for use during your learning process:

1) cue column; and

2) summary area

Now we will describe how you can take advantage of your new way of taking notes by adopting a systematic process for learning from them. The goal of this process is to increase your understanding of what was covered in class and to move as much as possible of what was covered from your short term memory to your long term memory through repetition, review, and reinforcement.

The process of learning from your notes can be described in six separate but interrelated steps.

Step 1. - Study and annotate your notes. Study each page of your notes and fill in any missing information. Add words, phrases, facts, or steps in a derivation you may have skipped or missed, and fix any difficult-to-decipher jottings. As you read your notes, bring them "alive" by making liberal use of colored pens or pencils. Highlight important points by underlining, circling or boxing, or using arrows.

Step 2 – Question or Reduce. Formulate a question answered by each major term or point in your notes and write it in the cue column. This is a little bit like the game TV show *Jeopardy* where the contestants are given the answer and asked to supply the question. An alternative approach would be to reduce each main idea or set of facts into a key word or phrase and write it in the cue column.

Step 3 – Summarize. Write a summary of each page in the summary area at the bottom. Summarizing forces you to think about the broader picture. Your summary should answer the question: "What is this page about?" These summaries will be particularly helpful in finding key information when you are studying for an exam.

Step 4 – Recite. Once you have studied and annotated your notes and filled in the cue column and the summary area, it is time for the most important step in your learning process—recitation. The process of reciting is relatively straightforward. Go back to the first page and cover the note-taking area with a blank sheet. Read the first question or key word in the cue column. If you wrote questions, answer each question in your own words. If you wrote key words, describe the main idea or set of facts referred to. Ideally, you should recite out loud. If you are reluctant or unable to recite aloud, recite by writing out your answers. Slide the blank sheet down to check your answer. If your answer is wrong or incomplete, try again. Continue this process until you have gone all the way through your notes.

Step 5 – Reflection. After you have completed the first four steps above, take some time to reflect on what you have learned. Ask yourself questions like: What's the significance of what I have learned? How do the main ideas of this lecture fit together into a "bigger picture"? How do they fit into what I already know? What are some possible applications of the key ideas from this lecture? Which ideas are clear? Which ideas are confusing? What new questions do the ideas raise?

Step 6 – Review. Working through the process described above will not only increase the amount you learn from your lectures and from your notes, it will also convert your lecture notes to study notes that will be useful in future reviews. My suggestion is that you review all of your notes once each week. Doing so won't take much time and will payoff immensely in the long term. You'll find that if you spend just 10 minutes in a quick weekly review of your notes, you'll retain most of what you learned initially. And then give your notes a more thorough review as you prepare for a test and then again as you prepare for the final exam. And don't forget to use reciting to verify what you know during each review process.

READING/REREADING THE TEXT. Next read or "reread" (if you read the text in preparation for the lecture) the text material. Follow the "reading for comprehension" methodology described in Section 5.1. Don't make the mistake of skimming over the material in order to get on to the assigned problems. Read to understand the concepts. Make reciting a key part of your reading process.

SOLVE PROBLEMS. As we previously discussed, solving one or two problems, even if that's all your professor assigns, will not ensure an adequate level of understanding. If time permits, work all of the problems in the book. If more time is available, work them a second time. Practice, practice, practice! The more problems you solve, the more you will learn.

> ***Much of the learning in math, science, and engineering courses comes not from studying or reading but from solving problems.***

To the extent possible, utilize the analytical problem solving methodology described in Section 5.1. By doing so, you'll improve your problem solving capability over time.

After you have gained a certain level of mastery of the material, you can reinforce your understanding through a group study session or by going to see your instructor during office hours to address specific questions or problems.

Only then will you be ready for the next class meeting. You will have reinforced your understanding of the material several times. Later you will again reinforce it when you review for a test and still later when you prepare for the final exam.

LEARN TO MANAGE YOUR TIME

Time is an "equal opportunity" resource. All people—regardless of their socioeconomic status, gender, ethnicity, physical challenges, cultural practices, or any other kind of "difference"—have exactly the same amount of time. Everyone, including you, gets 168 hours each week—no more, no less.

> ***There is no point in saying that you have no time, because you have just as much as anyone else.***

Time is an unusual and puzzling concept. Even the most brilliant scientists and philosophers aren't sure how to explain it. But we do know some things about it. It can't be saved. When it's gone, it's gone. It also seems to pass at varying speeds—sometimes too slowly and other times too quickly. And it can be put to good use, or it can be wasted. Some people accomplish a great deal with their time, while others accomplish virtually nothing with theirs.

People who accomplish a great deal, without exception, do two things:

(1) They place a high value on their time.

(2) They have a system for scheduling and managing their time.

Some of these systems are very sophisticated, and you may wish to look into acquiring one, particularly when you become a practicing engineer. As a student, you can do quite well with a long-term calendar to record your future appointments and with a simple form for making a detailed schedule of day-to-day plans for any given week (see form on page 200).

<u>HOW MANY HOURS SHOULD YOU STUDY?</u> Once you commit to staying on top of your classes and reinforcing your learning as often as possible, you must make sure that you are allotting a sufficient number of study hours to truly master the material covered in a one-hour lecture. Earlier, in presenting the "60-Hour Rule," we mentioned the standard rule-of-thumb that you should study two hours out of class for every hour in class. But this is often a gross oversimplification or, at best, a very limited generalization. In actual fact, the amount of study time required will vary from course to course, depending on such factors as:

- How difficult the course is
- How good a student you are
- How well prepared you are for the course
- What grade you want to receive

For demanding technical courses, it is doubtful that two hours of studying for every hour spent in class are enough. The appropriate number for you may be three, four, or even five hours. Although this may be difficult to assess, especially early on in your education, it's good to determine a number for each of your classes. You can always adjust it later.

> **REFLECTION**
>
> On a scale of 0-10, how good a student are you? Reflect on each of your courses. How difficult is the course (very difficult, moderately difficult, not difficult)? How well prepared are you for the course (very well prepared, fairly well prepared, not well prepared)? What grade do you want in the course (*A, B, C*)? Based on this information for each of your courses, write down how many hours you need to study for each one hour of class time.

Once you have decided that for a particular course you should study, say, three hours between one class meeting and the next, and you have blocked out a schedule for studying as soon after each lecture as possible, you have done the easy part. The hard part is actually doing it. Putting these approaches into practice requires you to be organized and skilled in managing your time.

<u>MAKING UP YOUR WEEKLY SCHEDULE</u>. Your effectiveness and productivity as a student will be greatly enhanced by scheduling your time. The approach I took when I was a student was to sit down each Sunday night with a form like the one shown on page 200 at the end of this chapter and schedule my entire week. You may find that a whole week is too much, and prefer to schedule a day or two at a time. That's fine. The idea is to find a scheduling method that works for you.

For whatever time period you choose, first write down all your commitments: classes, meetings, part-time work, time to get to and from school, time for meals, and so forth. The rest of the time is available for one of two purposes: study or recreation.

Next, schedule blocks of time to study. You have already decided how much time you need between one class meeting and the next, and you know the advantages of scheduling this time as soon after each class meeting as possible. Write down both where and what you will study. Students tend to waste too much time between classes making three decisions:

(1) Should I study now or later?

(2) Where should I study?

(3) What should I study?

By making these decisions in advance, you will eliminate this unnecessary waste of time.

REFLECTION

Do you schedule your study time? If so, how is it working for you? If not, why don't you? Sit down with a form like the one on Page 200 of this chapter. Schedule your time for the next week. After scheduling your commitments (class, work, appointments, etc), schedule your study time following the principles presented in this section. Include information on both where you plan to do the studying and what course you plan to study in each time block. Make a commitment to follow your schedule for the next week.

Once your study time is scheduled, check to see that you've left open some time for breaks, recreation, or "down time." If not, you are probably over-committed. You have taken on too much. One of the advantages of making a schedule is that it gives you a graphic picture of your situation. Remember, don't "program yourself for failure." Be realistic about what you can handle. If you are over-committed, you should probably let something go. Reduce your work hours or your extracurricular activities, or reduce the number of units you are taking.

MAKE A SERIOUS COMMITMENT TO YOUR SCHEDULED STUDY TIME. Making up a weekly schedule, you will find, is easy and fun. But sticking to it will be a challenge. The key is to make a serious commitment to your study time. I'm sure you take your class time as a serious commitment. If, for example, five minutes before a class a friend asked you to go have a cup of coffee or a coke, you would say, "Sorry. I can't because I have a class." But what about your study time? What if the same friend came up to you just as you were about to go to the library to study?

You need to make the same commitment to your scheduled study time as you do to your class time. After all, much more learning occurs out of class than in. It always astonishes me that students are so willing to negotiate away their study time. Every time you put off an hour of studying, you are giving up time that you cannot recapture, and that means borrowing time from the future. If, however, your future is already scheduled, as it should be, the notion of borrowing time from the future is impossible. You're talking about time that isn't there.

To monitor yourself, outline the hours you actually study in red on your schedule form. At the end of each week, you will be able to readily count up how much studying you did. If you are doing poorly in your classes, I'll bet you will see a direct correlation between your performance and the amount of studying you are doing.

Initially you may find that you have made a schedule you are unable to follow. Don't "beat yourself up" over that. And don't use it as an excuse to give up scheduling your time completely. Over time, you will learn about what you can and cannot do, and become more proficient at scheduling your time.

If you are like most students, you will find that by scheduling your time and following the schedule, you will feel as though you have more time than you did before. And your stress level will go down. Many students spend more time worrying about the fact that they are not studying than they do actually studying. "Tending to business" can give you a real sense of well being.

In summary, the benefits of scheduling your study time are:

- You will be able to see immediately if you are overextended.
- You are more likely to keep up in your classes and to devote adequate time to studying.
- You'll get immediate feedback as to how much you are actually studying.
- You'll learn about yourself—both what you can and can’t do.
- You'll feel that you have more time than you ever had before.
- You'll feel much less stressed-out over school.

DAILY PLANNING--"TO DO" LIST. One final approach to getting the most out of each day is to make up a daily "to do" list. To do this, take a few minutes each evening and write down a specific list of what you want to get done in the next 24 hours.

Then prioritize the items on the list. Either rank the items from top to bottom, or classify each as "high," "medium," or "low" priority. The next day, work on the most important items first. Try to avoid the urge to work on items that are easy or fun, but are of low priority. As you complete items, cross them off of your "to do" list. At the end of the day, evaluate your progress and reschedule any items that remain on your list. Once again, though, if you repeatedly find that you can't accomplish everything on the list, you are probably over-scheduling yourself. And having to reschedule unaccomplished "to do" items means borrowing from the future, time that isn't there.

USING A LONG-TERM PLANNER. In addition to planning each week, you need a way to keep track of long-term commitments, important dates, and deadlines. Your campus bookstore or a local office supply store has both academic year planners and calendar year planners for this purpose.

Enter appointments, activities, events, tasks, and other commitments that extend beyond the current week in this planner. These might be academically-related, such as test dates, due dates for laboratory reports or term papers, meetings of student organizations, engineering seminars or guest speakers, and advising appointments. Also include personal appointments such as medical and dental checkups and car maintenance schedules; special occasions such as birthdays, anniversaries, and holidays; and recreational activities such as parties, concerts, plays, and other cultural events.

Each week, as you make up your weekly schedule, transfer commitments from your long-term planner to your weekly schedule.

You may consider keeping your weekly schedules and long-term planners so that in the years to come you can enjoy them as a reminder of what you did during this uniquely important period of your life.

PRIORITY MANAGEMENT.

If you want to move to a higher level of managing your life and your affairs, Stephen Covey in his powerful book *Seven Habits of Highly Effective People* [2] points the way. Covey's guiding principle is to:

Organize and execute around priorities.

Priority management means doing what needs to be done. There are two dimensions to deciding what needs to be done:

- How urgent is it? (Requires immediate attention; or doesn't require immediate attention)
- How important is it based on personal values? (Important; or not important)

These two dimensions—urgency and importance—should not be confused, but frequently are. It's almost second nature to think that anything that's urgent must be important. The phone rings, we have to answer it. Our favorite TV show is on, we have to watch it. A friend wants to talk, we have to talk. Urgent matters press on us. They insist on our attention. They're often popular to others. And often they are pleasant, easy, fun to do. But many urgent matters are not important!

"Importance" relates to whether it needs to be done at all. Not important should mean we don't do it at all. Much of our time and effort is devoted to tasks that are not important, whether they are urgent or not.

These dimensions can be shown visually by the following 2x2 matrix having four quadrants:

I **Urgent** **Important**	**II** **Not urgent** **Important**
III **Urgent** **Not Important**	**IV** **Not urgent** **Not Important**

Key to the process of priority management is the criteria we use to determine what is important. This depends on our value system. Suffice to say, candidates for high personal value include:

School
Family

Friends
Health
Personal goals

Stay out of Quadrants III and IV. People who spend time almost exclusively in Quadrants III and IV lead basically irresponsible lives. Effective people stay out of Quadrants III and IV because, urgent or not, activities in these quadrants aren't important. You might have already guessed that staying out of Quadrant III will require you to become good at saying "no."

Activities in Quadrant I can be described as what's called "crisis management" activities. Much of your life is dominated by activities that are both urgent and important. Time to go to class. Got to prepare for tomorrow's exam. Term paper due. Got to go to work. All important and urgent things.

We can't ignore the urgent and important activities of Quadrant I. However, our overall effectiveness will be controlled by Quadrant II—i.e., how we do with the things that are important but don't have to be done today. Since we can't skip Quadrant I activities, finding time for Quadrant II activities will require that we give up activities from Quadrants III and IV. One bit of good news. In time, choosing Quadrant II activities will have the benefit of reducing the need to always operate from the crisis management perspective of Quadrant I.

REFLECTION

Reflect on the academic success strategies presented thus far in the chapters indicated:

Structuring your life situation (Chapter 1)
Preparing for lectures (Chapter 4)
Seeking one-on-one instruction from your professors (Chapter 4)
Utilizing tutors and other academic resources (Chapter 4)
Scheduling your study time (Chapter 5)
Mastering the material presented in each class before the next class comes (Chapter 5)

Are these Quadrant I, II, III, or IV activities? Do they have to be done immediately (urgent or not urgent)? Are they important? Or not important?

5.3 PREPARING FOR AND TAKING TESTS

As you learned in Chapter 1, a vital component of successful engineering study is becoming a *master* at preparing for and taking tests.

PREPARING FOR TESTS

Clearly, the best way to prepare for tests is to practice the many strategies discussed earlier. When I hear a student boast that he or she stayed up all night studying for a test, I know this is a student who is not doing well. You, too, should recognize this by now. This is a student who most likely does not study from class to class, does not schedule his or her time well, does not understand the learning process (i.e., the need for incremental, reinforced learning), and does not realize the pitfalls of studying alone (the image of a student staying up all night studying for a test certainly fits the "lone wolf" metaphor, doesn't it?).

The truth is, if you have incorporated the study skills we have discussed into your regular study habits—even just the one skill of "taking it as it comes"—preparing for a test is not very hard. It merely involves adjusting your schedule several days prior to the test to review the material. You should never have to cover new material when preparing for a test.

There is, however, one major aspect of test taking that distinguishes it from all other forms of studying and learning: time pressure. To do your best on tests, therefore, you need to learn how to work under the pressure of time.

Here are some useful tips that will both improve your performance on tests and lessen your anxiety about taking them. Several days before a test, spend a portion of your study time working problems under a time limit. If you can, obtain tests from previous semesters or, better yet, construct your own. Creating and taking your own practice exams will give you invaluable experience in solving problems under pressure, plus it will give you the added advantage of learning to "scope out" tests. In time you will significantly improve your ability both to work under pressure and to predict what will be on tests.

Unlike the student who stays up all night frantically cramming, be sure to get a good night of sleep before a test. Arrive at the test site early so you have ample time to gather your thoughts, and be sure you have whatever materials you'll need: paper, pencils, calculator. A certain amount of "psyching yourself up," similar to what an athlete does prior to a big game, might be helpful; however, you don't want to get so nervous that you can't concentrate.

TEST-TAKING STRATEGIES

When you are given the test, don't start work immediately. Glance over the entire test first, and quickly separate out the easier problems from the harder ones. Many instructors grade on a "curve," which means that your grade will be based on its relation to the class's average performance, not your individual score alone. If this is the case, you also need to size up the overall difficulty of the test and make a guess as to what the class average will be. In fact, jot down your estimate so that you can compare it later with the actual outcome. Through this process, over time you will become adept at sizing up tests. You will be able to recognize that on one test, it may take a score of 90 to get an *A*, while on another test it may only require 50. Knowing that you only need to get a portion of the problems correct for a good grade will greatly affect the way you approach a test.

Once you have sized up the test, don't start with the first problem; start with the easiest one. As you work the easier problems and accumulate points, your confidence will build and you will develop a certain momentum. But always keep an eye on the clock. If you divide the time available by the number of problems, you will know approximately how much time to spend on each. Use this as a guide to pace yourself. Also, try to complete a problem before leaving it, and avoid jumping from one uncompleted problem to another, since you will waste time getting restarted on each.

Although you are under a time constraint, be sure to work carefully and attentively, as careless mistakes can be very costly. It is probably smarter to work three of five problems carefully than to do all five carelessly. And by all means, never leave a test early. What do you have to do that could be more important than achieving the highest possible score on a test? If you have extra time, check and recheck your work. No matter how many times you proofread a term paper, mistakes can still be overlooked. The same is true for a test.

5.4 MAKING EFFECTIVE USE OF YOUR PEERS

We close this chapter, with one of the most important academic success strategies—making effective use of your peers. Your peers can significantly influence your academic performance, either positively or negatively.

Negative peer pressure put on those who apply themselves to learning is an age-old problem. Derisive terms like *dork*, *wimp, nerd*, *geek*, and *bookworm* are but a few of those used to exert social pressure on the serious student. You may have experienced this type of peer pressure in high school if your friends were not so serious about their academics as you, and you may have been forced into a pattern of studying alone—separating your academic life from your social life.

> ***The "lone-wolf" approach to your academics may have worked for you in high school, but it is doubtful that it will work for you in engineering study where the concepts are much more complex and the pace much faster.***

Even if you are able to make it through engineering study on your own, you will miss out on many of the benefits of collaborative learning and group study.

Overview of Collaborative Learning

In a previous section, we discussed teaching modes: large lectures, small lectures, recitations, tutoring sessions. Now we turn to ***learning modes***. There are really only two:

(1) Solitary

(2) Collaborative

Either you try to learn by yourself, or you do it with others.

As I travel the country, I always make a special effort to visit Introduction to Engineering classes, where I make it a point to ask students, "How many of you when you study spend some fraction of your study time on a regular basis studying with at least one other student?" Generally, in a class of 30 students, three or four hands will go up. Then I ask, "How many of you spend virtually 100 percent of your study time studying by yourself?" And the remaining 90 percent of hands go up.

> ***My anecdotal research indicates that about 90 percent of first-year engineering students do virtually 100 percent of their studying alone.***

Hence, the predominant learning mode in engineering involves a student working alone to master what are often difficult, complex concepts and principles, and then apply them to solve equally difficult, complex problems.

The fact that most students study alone is indeed unfortunate because research shows that students who engage in collaborative learning and group study perform better academically, persist longer, improve their communication skills, feel better about their educational experience, and have enhanced self-esteem. We just read essentially the same message in that excerpt from the Harvard University study a few pages back. As even more evidence, Karl A. Smith, Civil Engineering professor at the University of Minnesota and a nationally recognized expert on cooperative learning, has found that [3]:

Cooperation among students typically results in:

a. Higher achievement and greater productivity

b. More caring, supportive, and committed relationships

c. Greater psychological health, social competence, and self-esteem

In my own anecdotal research, I have tried to understand why students study alone, so I also make it a point to ask students, "Why don't you study with other students?" I almost always get one of these three answers:

(1) "I learn more studying by myself."

(2) "I don't have anyone to study with."

(3) "It's not right. You're supposed to do your own work."

The first of these reasons is simply wrong. It contradicts all the research that has been done on student success and student learning. The second reason is really an excuse. Your classes are overflowing with other students who are working on the same homework assignments and preparing for the same tests that you are. The third reason is either a carryover from a former era when the culture of engineering education emphasized "competition" over "collaboration," or it comes from that romanticized ideal of the "rugged individualist" that we debunked in Chapter 3. Today, the corporate buzzwords are "collaboration" and "teamwork," and engineering programs are under a strong mandate to turn out graduates who have the skills to work well in teams.

If you are using any of these reasons to justify your "lone-wolf" approach to your academic work, you should now see their inherent problems and, thus, you need to change your approach.

The true value of academic relationships is illustrated in this cartoon:

Campus radical.

BENEFITS OF GROUP STUDY

If you're still not convinced, then look at the issue from a different perspective. Instead of focusing on the weaknesses or problems of solitary study, consider the strengths or benefits of group study. In this new light, you will find three very powerful and persuasive reasons for choosing the collaborative approach over the solitary one:

(1) You'll be better prepared for the engineering "work world."

(2) You'll learn more.

(3) You'll enjoy it more.

Each of these is discussed in the following sections.

YOU'LL BE BETTER PREPARED FOR THE ENGINEERING WORK WORLD. Whether you choose to study alone or with others often depends on your view of the purpose of an engineering education. If you think the purpose of that education is to develop your proficiency at sitting alone mastering knowledge and applying that knowledge to the solution of problems, then that's what you should do. However, I doubt you will find anyone who will hire you to do that. It's not what practicing engineers do by and large.

So if you spend your four or five years of engineering study sitting alone mastering knowledge and applying that knowledge to the solution of problems (and perhaps becoming very good at it!), you will have missed out on much of what a quality education should entail.

A quality education provides you not only the ability to learn and to apply what you learn, but also the ability to communicate what you have learned to others; the ability to explain your ideas to others and to listen to others explain their ideas to you; and the ability to engage in dialogues and discussions on problem formulations and solutions. You may land on a very important "breakthrough" idea, but if you can't convince others of it, it is unlikely that your idea will be adopted.

YOU'LL LEARN MORE. Do you recall our earlier discussion of traditional teaching modes, all of which kept learning to a minimum? In essence, group study and collaborative learning take up where the traditional modes leave off—and the result is an increase in what you learn.

There are a number of ways to explain how this happens. One is the adage that *"two minds are better than one."* Through collaborative study, not only will more information be brought to bear, but you will have the opportunity to see others' thought processes at work. Perhaps you have played the game *Trivial Pursuit*. It always amazes me how a small group of people working together can come up with the answer to a question that no member of the group working alone could have done.

Another explanation comes from the claim that:

If you really want to learn a subject, teach it.

It's true! As an undergraduate engineering student, I took three courses in thermodynamics. Yet I didn't really understand the subject until I first taught it. When two students work together collaboratively, in effect, half the time one student is teaching the other, and half the time the roles are reversed.

YOU'LL ENJOY IT MORE. Group study is more fun and more stimulating than solitary study, and because you'll enjoy it more, you are likely to do more of it. This wonderful benefit of group study can be illustrated by the following personal story.

My Own Experience with Group Study

When I was working on my Ph.D., a close friend of mine and I took most of our courses together. To prepare for exams, typically we would meet early on a Saturday morning in an empty classroom and take turns at the chalkboard deriving results, discussing concepts, and working problems. Before we knew it, eight or ten hours would have passed. There is no way I would have spent that amount of time studying alone on a Saturday at home. Would you? The temptations of TV, the Internet, telephones, e-mail, and friends, along with the need to run errands or do work around the house, would surely have prevailed over my planned study time. By integrating my academic work with my social needs, I enjoyed studying more and did more of it.

FREQUENTLY ASKED QUESTIONS ABOUT COLLABORATIVE LEARNING

Once students embrace the concept of collaborative learning, they generally have questions on how to make it work. The three most frequently asked (and probably most important) questions are:

- What percentage of your studying should be done in groups?
- What is the ideal size of a study group?
- What can be done to keep the group from getting off task?

Although there are no definitive answers to these questions, the following points serve as fairly reliable guidelines.

PERCENTAGE OF TIME. Certainly, you should not spend all of your study time working collaboratively. I would suggest somewhere between 25 and 50 percent. Prior to coming together, each member of a group should study the material and work as many problems as possible to gain a base level of proficiency. The purpose of the group work should be to reinforce and deepen that base level of understanding. The better prepared group members are when they come together, the more they can accomplish during their study sessions.

Size of Study Group. When you hear the term "study group," what size group do you think of? Five? Ten? Fifteen? My ideal size is ***two***. Study "partners." When two people work together, it is easier to maintain a balanced dialogue, in which each is the "teacher" for half the time.

Triads can work well too. In larger groups, however, it can be difficult to ensure equal participation, and members often feel the need to compete for their "fair share" of the time. Even between study partners, a conscious effort may be required to keep one of the two from dominating the dialogue. So my advice is to keep the groups small. If more people come together to study, it's okay. Generally, subgroups of twos or threes will develop.

But the idea that two is the ideal size of a study group is my personal preference. Richard Felder of North Carolina State University sent me the following different view:

> "With two people you don't get sufficient diversity of ideas and approaches, and there's no built-in mechanism for conflict resolution, so the dominant member of the pair will win most of the debates, whether he/she is right or wrong. Five is too many—someone will usually get left out. I suggest three as the ideal size, with four in second place, and two in third."

I would encourage you to experiment to see what works best for you.

Staying on Task. You may find it difficult to stay on task when working with others. There are no simple solutions to this problem, for it really boils down to each student's discipline and commitment to his or her education. Once again, though, size may be a factor: the larger the group, the more difficult it will be to keep everyone focused on academics. Yet even in groups of two or three, staying on task can be a problem.

I have found it helpful to split up a group's meeting time into a series of short study periods with breaks between each period. Agree, for example, to study for 45 minutes, and then take a 15-minute break. After the break, it's back to work for another 45 minutes, followed by another 15-minute break. And so on.

If nothing else seems to help your group to stay on task, then you're left with only one solution: *just do it.*

REFLECTION

Do you spend some fraction of your study time on a regular basis studying with at least one other student? Or do you spend virtually 100 percent of your study time studying alone? If you don't study with other students, why not? Did the ideas presented in this section persuade you of the value of group study and collaborative learning? Are you willing to try it out? Make a commitment to identify a study partner in one of your key classes and schedule at least a two-hour study session with that person.

It Really Works

I often conduct workshops on collaborative learning, and at some point I have half of the class work on a problem in small groups and the other half work by themselves on the same problem. After about ten minutes, the ones who are working alone start looking at their watches and appear restless and bored. When time is called after 45 minutes, those who arc working in groups are disappointed and ask for more time. They often express that they are just getting "hot" on a solution to the problem.

Then, the next day I ask, "How many of you continued thinking about, working on, or talking to others about the problem we did yesterday?" Most of those who worked in groups raise their hands, whereas those who worked alone do not.

NEW PARADIGM

Collaboration and *cooperation* represent a major new paradigm shift in business and industry, replacing the paradigm of *competition* that began with the Industrial Revolution and held sway well into the 20th century.

Collaborative learning represents the same paradigm shift in engineering education. *Collaborative learning* is consistent with modern engineering management practice and with what industry representatives tell us they want in our engineering graduates. *Competition* and *individual achievement* are outdated notions, and rightly so. W. Edwards Deming, father of the "quality" movement, makes a compelling case [4]:

> "We have grown up in a climate of competition between people, teams, departments, divisions, pupils, schools, universities. We have been taught by economists that competition will solve our problems. Actually, competition, we see now, is destructive. It would be better if everyone would work together as a system, with the aim for everybody to win. What we need is cooperation and transformation to a new style of management . . . . Competition leads to loss. People pulling in opposite directions on a rope only exhaust themselves. They go nowhere. What we need is cooperation. Every example of cooperation is one of benefit and gains to them that cooperate. Cooperation is especially productive in a system well managed."

I hope you will be proactive in seeking opportunities for cooperation and collaboration with your fellow students, and that in doing so you will reap much greater rewards than you would have through competition and individual effort.

SUMMARY

This chapter addressed the important topic of "Making the Learning Process Work for You." Effective learning involves many skills and as in developing any skill, "practice makes perfect."

We began the chapter by discussing two very important learning skills—reading for comprehension; and analytical problem solving. Your success in engineering study will depend in large part on your skill in these two areas.

We then described the process of organizing your learning process. We emphasized perhaps the most important academic success strategy in the chapter, if not in the entire book—the need to keep up in your classes by mastering the material presented in each class before the next class comes. Approaches for mastering material and important time and priority management skills were also presented.

Then we discussed strategies for preparing for and taking tests. This subject deserves particular attention since most of your grade in math/science/engineering coursework will be based on your performance on in-class tests and exams.

The chapter concluded with approaches for making effective use of one most important resources available to you—your fellow students. By working collaboratively with your peers, particularly in informal study groups, sharing information with them, and developing habits of mutual support you will learn more and enjoy it more. At the same time, you will become well prepared for the engineering "work world," where teamwork and cooperation are highly valued.

On a concluding note, we want to stress that implementing the success strategies presented in this important chapter requires you to change—change how you think about things (your attitudes), and change how you go about these things (your behaviors). Thus, the value of any of the strategies mentioned above—indeed, the value of this entire book—depends on the extent to which you can make such changes. To help you succeed in what can be a difficult process, in Chapter 6 we talk about the psychology of change, offer ways for you to gain insights into yourself, and detail a step-by-step process to facilitate your personal growth and development.

REFERENCES

1. Polya, George, *How to Solve It: A New Aspect of Mathematical Method*, Princeton University Press, 2004.
2. Covey, Stephen R., *The Seven Habits of Highly Effective People*, Simon & Schuster, New York, NY, 1989.
3. Smith, Karl A., "Cooperation in the College Classroom," Notes prepared by Karl A. Smith, Department of Civil Engineering, University of Minnesota, Minneapolis, MN, 1993.
4. Deming, W. Edwards, *The New Economics for Industry, Government, Education*, MIT Center for Advanced Study, Cambridge, MA, 1993

PROBLEMS

1. Prepare a 10-minute talk on the methodology presented in Section 5.1 on *Reading for Comprehension*. Persuade a classmate from one of your other classes (not your Introduction to Engineering course) to listen to your talk and give you feedback on how well you understand and communicate the concepts.

2. Make a commitment to follow the steps in the methodology presented on *Reading for Comprehension* in Section 5.1 for one week. Be particularly attentive to the three "Before You Read" steps and to the recitation step "After You Read." Write a one-page paper on how the methodology impacted your learning.

3. Solve this problem following each of the four steps for analytical problem solving presented in Section 5.1. Use the problem solving strategies of *solving a simpler problem* and *drawing a diagram.*

   A painter built a ladder using 18 rungs. The rungs on the ladder were 5.7 inches apart and 1.1 inch thick. What is the distance from the bottom of the lowest rung to the top of the highest rung?

4. Solve this problem following each of the four steps for analytical problem solving presented in Section 5.1 using the *guess and check* strategy. Using the *make a table* strategy. Using the *solve an equation* strategy.

   Jose has 26 nickels and dimes in his piggy bank. The number of nickels is 2 fewer than three times the number of dimes. How much money does he have in his piggy bank?

5. Using the form presented at the end of this chapter, schedule your study time for one week. Attempt to follow the schedule. Write a one-page paper describing what happened.

6. Make a "To Do" list of things you need to do. Place each item in one of the four quadrants of Covey's matrix shown on Page 185. How many items are in Quadrant I (Urgent; Important)? How many are in Quadrant II (Not Urgent; Important)? How many are in Quadrant III and IV (Not Important)?

7. Go to your campus library and find a book on study skills. Check out the book and scan its Table of Contents. Identify three interesting sections and read them thoroughly. Then write an essay on why you picked the topics you did and what you learned about them.

8. If you studied for 100 hours, how many of those hours would be spent studying alone and how many would be spent studying with at least one other student?

9. If your answer to Problem 8 was that you spend most of your time studying alone, seek out a study partner in one of your math/science/ engineering classes. Get together for a study session. Write down what worked well and what didn't work well.

10. Interview two junior or senior engineering majors, and ask the following questions:

    a. What was the main difference they found between high school and engineering study?

    b. What were the most important new study skills they had to learn?

    c. What approach do they use to manage their time effectively?

    d. What do they think about group study? Its values and benefits?

	MONDAY	TUESDAY	WEDNESDAY	THURSDAY	FRIDAY	SATURDAY	SUNDAY
8-9							
9-10							
10-11							
11-12							
12-1							
1-2							
2-3							
3-4							
4-5							
5-6							
6-7							
7-8							
8-9							
9-10							

CHAPTER 6
Personal Growth and Development

You will either step forward into growth,
or you will step back into safety.

Abraham Maslow

INTRODUCTION

The focus of this chapter is your personal growth and development. Your success as a student and, later, as an engineering professional will depend on the extent you grow and develop both during and after college.

We begin the chapter by tapping into the paradigm of "continuous improvement" currently espoused by U.S. business and industry, and we urge you to adopt a personal plan of "continuous improvement" for every area you need to strengthen or change. We call this process *student development.*

To achieve the changes that your student development plan will entail, we then present a step-by-step process based on behavior modification theory.

We recognize that to change yourself, you must understand yourself. We therefore address three topics that conduce to self-understanding:

- Maslow's Hierarchy of Needs helps you understand your basic human needs, which must be met before you can concentrate on achieving your highest need *self-actualization.*
- A discussion of self-esteem shows you the "domino effect" that your personal development plan will have on your sense of self.

As you make changes in thoughts and actions that move you closer to your goals, you will feel increasingly better about yourself.

- A presentation about peoples' personality types will help you in both understanding yourself and understanding others.

We then go into more detail on the important topic of understanding others. Understanding others is in large measure an extension of the process of understanding yourself. Your success in your career will depend on your ability to work with people who are different from you—not only people having different learning styles and personality types, but also people who differ in their ethnicity, culture, and gender.

Next, we return to and expand upon personal assessment. We home in on ways to identify your strengths and areas for improvement so that your personal development plan is as thorough and up-to-date as possible. Within this context, we address two important (and often overlooked) areas for personal growth and development—communication skills and mental and physical wellness.

Finally, we leave you with several motivational messages to further strengthen your commitment to ongoing self-development. Your growth and development will result in increased success in your engineering education and innumerable successes in the years beyond college.

6.1 Personal Development - Receptiveness to Change

I usually start the first meeting of my Introduction to Engineering course by asking the students, "How many of you want to change something about yourself?" Generally, only three or four out of 30 students raise their hands.

This resistance to change, I think, has the same roots as students' reluctance to seek help (discussed in Chapter 3), in that both are seen by students as an admission that something is "wrong" with them.

This, we have already shown, is a counterproductive attitude, but not for students alone. Resistance to change was a powerful force in post-World War II business and industry practices in the United States, and a big factor in losing our #1 position in the world economy. While other countries such as Japan, Korea, Taiwan, and Germany were striving for "continuous improvement," we in the U.S. were satisfied with the status quo. Our motto for a long time was:

If it ain't broke, don't fix it.

Only recently, in the last 30 years or so, has U.S. industry changed its tune dramatically, replacing the "status quo" paradigm with one of "continuous improvement" to regain its competitive edge.

TOTAL QUALITY MANAGEMENT

The term that has become synonymous with "continuous improvement" in business and industry practices worldwide is "Total Quality Management," or simply "TQM." Developed in the 1950s and early '60s primarily by Japanese industrialists—along with significant input from noted American W. Edwards Deming—TQM espouses the philosophy that no matter how good we are, we should strive *continuously* to improve our quality. The true practitioners of TQM, therefore, do not attach shame or resistance to change. They are not only receptive to change; they actively seek it out.

As part of your engineering education, you will undoubtedly hear a lot about TQM (or its more modern counterpart *Six Sigma*). For now, though, a general overview of the TQM process will suffice. Like the engineering design process, the TQM process consists of a series of steps or stages. The first involves defining what "quality" means—a definition that will change from one context to the next, depending on who the customer is and what the customer's needs are. In the second stage, performance measures (*metrics*) are established to meet and, if possible, exceed the customer's needs. Last, a detailed plan is drawn up and implemented.

"PERSONAL" TOTAL QUALITY MANAGEMENT

I hope I can persuade you to adopt a ***personal TQM philosophy***. The "customer" can be you or someone else, such as your parents, your spouse or partner, your professors, or your future employer. Regardless of whom you choose, what's important is that you strive to change, grow, and improve yourself *continuously* in every area that impacts your effectiveness in meeting and exceeding the needs and expectations of your customer. Your motto should be:

Even if it ain't broke, try to improve it.

Consider, for example, a major league baseball player whose batting average is .315. This person was a star in high school, a star in college, and now a superstar in the major leagues. He makes $12 million a year. Yet he still works two hours a day with his batting coach to raise his average to .320. In fact, the reason his batting average is .315 is that he wasn't satisfied when it was .295.

The basic message is that people who are successful recognize the need to strive continuously to change, grow, and improve. Wanting to improve has nothing to do with the idea that there is something "wrong" with you.

REFLECTION

Reflect on the idea of growing, changing, and improving. Do you embrace the idea? What would you like to change about yourself? Become a better writer? Improve your "people skills"? Stop procrastinating? Learn to cook? Feel more comfortable in your own skin?

STUDENT DEVELOPMENT

When I talk about a personal TQM philosophy, I prefer to phrase it as *student development*, and I tailor both the TQM process and vocabulary to make them more student-oriented. The general concept of "continuous improvement," however, remains unchanged.

The cornerstone of student development is well-defined goals—the counterpart of TQM's first step of defining a customer's needs. As a student, your immediate goal is earning your B.S. degree in engineering. That's a given. You may have other goals as well—achieving a certain grade point average, finding the job you want, and performing well in that job. Over the long term, you may want to have a successful career as a practicing engineer or become president of your own company.

In previous chapters, we have talked extensively about the importance of goals. With regard to student development, they play yet other roles: they specify the areas in which you need to grow, change, or develop; and they provide the necessary foundation for tracking the progress of your personal growth (the counterpart of TQM's second step of defining performance *metrics*). Only with clear goals can you—or anyone else—make value judgments about your behavior. Let me give you an example.

If a student were to tell me, "My friends have invited me to go to Las Vegas this weekend," I would have no way of assigning a value judgment to her statement. If, however, she added, "I probably won't go, because I really want to graduate this year, and I'm a little behind in my thermodynamics class," a value judgment would be easy to make, and I would support her decision not to go.

On the other hand, if the same student explained, "I just finished my final exams. I studied really hard and did great," I would be inclined to say, "Congratulations! Now you're only a semester away from graduating. So have a great time in Vegas."

In this example, it was not until I knew the student's goal *("I really want to graduate this year.")* that I could place a value judgment on her proposed weekend trip. Our goals, then, provide the context we need to assess what we do—or propose to do.

VALUE JUDGMENTS APPLIED TO OUR ACTIONS, THOUGHTS, AND FEELINGS

In forthcoming sections, we will examine in depth how our goals and value judgments fit into a larger process of change and personal growth. As a preface to this examination, let's look more closely at valuc judgments—the *metrics* of our personal TQM program—and the behaviors to which they apply.

In the example I just gave, my value judgments pertained to the student's actions (or proposed actions). **Actions**, what we say and do, are one part of human behavior. We also have **thoughts**, the ideas or attitudes we hold; and **feelings**, the emotions we have. Obviously, our actions, thoughts, and feelings are deeply interrelated: our feelings can affect our thoughts and actions, our actions can impact our thoughts and feelings, and so forth. But it is helpful to separate them out when we are talking about personal (i.e., student) development, for doing so establishes a framework for analyzing, understanding, and changing ourselves.

Analyzing our actions, thoughts, and feelings leads to the value judgments we make about each. For this purpose, we can classify our actions in one of two ways: **productive** and **non-productive**.

Productive actions support the achievement of our goals.

Non-productive actions do the opposite: they interfere with or work against the achievement of our goals.

Similarly, we can classify our thoughts as either **positive** or **negative**.

Positive thoughts result in our choosing productive actions.

Negative thoughts result in our choosing non-productive actions.

If we return to the example on the previous page, we can see how our thoughts can affect our actions in either positive or negative ways. When the student in the example announced, *"My friends have invited me to go to Las Vegas for the weekend,"* she proposed an **action**. Before acting on the proposal, however, she voiced two **positive thoughts**: *"I really want to graduate this year"* and *"I probably won't go,"* which led to a **productive action**—i.e., staying home to catch up in her thermodynamics class.

In this case, we see positive thoughts conducing to a productive action. The reverse can also happen, as the following examples of negative thoughts leading to non-productive actions show:

Negative Thought	Non-Productive Action
"I'm so far behind, I don't get anything out of going to class."	Cut class.
"I learn better studying by myself."	Spend 100 percent of study time studying alone.
"Physics is too hard. I just can't do it."	Procrastinate; put off studying.
"Professors don't seem to want to help me. They make me feel stupid."	Avoid seeking help from professors outside of class.
"I don't like having my life run by a schedule."	Waste time by not scheduling it.
"I don't have time for student organizations."	Avoid participation in student organizations.
"I'm not good at writing and don't like doing it."	Avoid opportunities to develop writing skills.

Finally, our **feelings** can be classified as either **positive** or **negative**.

Positive feelings produce positive thoughts, which in turn lead to productive actions.

Negative feelings produce negative thoughts, which lead to non-productive actions.

Sources of feelings, particularly negative ones, are not always easy to pinpoint. Because they are often connected to our self-esteem, they may be hidden, locked away in our unconscious minds as nature's way of "protecting" us from dangerous or unpleasant experiences. If this is the case, it normally takes time and a concerted effort with the help of a therapist or counselor to uncover them.

Self-esteem and the consequent feelings we have—conscious or unconscious—are extremely important factors in the student development process. We therefore will devote a separate section later in the chapter to discuss these factors in depth. For now, it is enough for you to know that feelings constitute a distinct part of being human and, like our thoughts, they can be judged as either positive or negative. We should also acknowledge that most of our feelings are apparent. Whenever we describe ourselves as happy, sad, excited, ashamed, affectionate, angry, comfortable, guilty, pleased, nervous—or any of the many other words that refer to our emotional state, we are expressing our feelings.

To give you just one example of how our feelings can affect our thoughts and actions, read the following story about "Jane R."

Jane R. gets terrible feelings of anxiety when she has to speak in public. She has thoughts like, "I'm a lousy speaker." When she performs the action, she gets so nervous that she does a poor job.

Out of desperation, she goes for counseling. During therapy, she recalls that in elementary school, she was criticized by her teacher when called on to read aloud to the class. This experience left her traumatized.

As an adult, she is able to re-examine that experience and realize that it was okay for a third-grader to make a mistake when reading aloud. She also realizes that her teacher didn't mean to hurt her. As a result, she is able to forgive herself and her teacher. By diffusing the negative feelings, she begins to think, "Maybe I can do a good job of speaking in public." And she does!

If we were to analyze the cause-effect relationships of Jane R.'s feelings, thoughts, and actions—first before, and then after her therapy—the process could be diagrammed as follows:

BEFORE THERAPY

Negative Feelings →	Negative Thoughts →	Non-Productive Action
Very anxious prior to making presentations	"I'm a lousy public speaker."	Does a poor job when making a presentation

AFTER THERAPY

Positive Feelings →	Positive Thoughts →	Productive Action
"Okay," even optimistic about future presentations	"Maybe I can do a good job of speaking in public."	Performs well on subsequent presentations

REFLECTION

Reflect on your recent actions. What are examples of actions you took that were supportive of your goal of becoming an engineer (productive actions)? What thoughts led you to take these actions? What are examples of actions you took that worked against or interfered with your goal of becoming an engineer (non-productive actions)? What thoughts led you to take those actions?

THERAPY AND COUNSELING AS CHANGE AGENTS

Jane R.'s story calls attention to the potential value of *counseling* and *therapy* in the process of personal development. Although it typically requires a costly and lengthy commitment, psychotherapy is sometimes the only way individuals can change themselves in positive ways. It was Jane. R.'s only solution. It may be yours, too.

Jane R.'s case also illustrates the basic premise of counseling or therapy, which is that any ongoing, unresolved negative feelings likely resulted from some traumatic childhood experience, which has been buried away in the unconscious mind. If this is true, the job of the psychotherapist is to dig beneath one's surface feelings to find their root cause. Once uprooted, the originating source and feelings generally lose the cathectic force they were able to exert over the years, and the individual is then able to deal with them on a conscious, rational level—or as Jane R. did, discard them entirely.

This is, perforce, a cursory overview of the counseling process, but I hope it gives you a basic understanding of how therapy works and when it is useful to overcome certain obstacles standing in the way of your personal growth. It assumes that human behavioral change must start with one's feelings, turning negative feelings into positive ones—a process in itself that can be very difficult, very time-consuming, very costly, but also very beneficial. Only then can new positive feelings lead to positive thoughts and, finally, to productive actions.

BEHAVIOR MODIFICATION AS A PROCESS FOR CHANGE

Behavior modification is another effective mechanism for changing your actions, thoughts, and feelings. For students, it's probably the most accessible and practical approach if you are truly committed to personal growth and change.

The premise of behavior modification is somewhat different than that of counseling/therapy. It assumes that human behavioral change should start "at the top" with your actions, which you consciously choose. If these actions are non-productive, you have the option to replace them with productive actions, and so instigate a process that filters "down" to your thoughts and feelings.

According to behavior modification theory, you have less control over your thoughts. You cannot help having negative thoughts. However, you can become more conscious of them and so try to change them to positive thoughts. Generally, you can do this by finding a higher context for your thinking. Once again, your goal will provide that context.

An Example

You have a test coming up in your math class. A productive action would be to study from 7:00 to 10:00 p.m. tonight. Behavior modification would hold that you are completely capable of choosing that action. However, you may have thoughts such as "I'd rather go out with my friends" or "I'm tired and don't feel like studying tonight." These are negative thoughts because, if acted on, they will lead you to a non-productive action (i.e., not studying).

Your challenge is to recognize that such thoughts are negative and try to change them. For example, the thought that "I don't feel like studying tonight" can be changed to "I really do want to study tonight because doing well in my math class will move me closer to getting my engineering degree."

As already noted, you have much less direct control over your feelings, because they are often tied to your self-esteem. While the issue of self-esteem will be dealt with in a subsequent section, you saw roughly how counseling/therapy starts with one's feelings to enact changes in one's behaviors. A similar rough sketch can be drawn to show how behavior modification works:

> ***If you begin to choose productive actions in support of a personal goal, and if you work to change negative thoughts to positive ones in support of those actions, in time you will feel more positive about yourself and about your life.***

In summary, the ***Student Success Model*** on the next page shows how behavior modification—or human behavioral change in general—works.

As illustrated there, achieving your goal of graduating in engineering requires that you change your actions from non-productive to productive ones, your thoughts from negative to positive, and (to the extent possible) your feelings from negative to positive ones. Through these changes you will grow and develop as a student.

6.2 Making Behavior Modification Work for You: Three Steps to Achieve Change

Sound easy? Just change those negative thoughts to positive ones; start choosing productive actions; and everything will be just great. Actually, you will find that change is not so easy. Changing your behavior requires you to successfully navigate three steps, each of which can offer significant barriers to change:

Step 1. Knowledge - "You know what to do."
Step 2. Commitment - "You want to do it."
Step 3. Implementation - "You do it."

Step 1. Knowledge—"You know what to do."

By *knowledge*, we mean that:

You know what to do.

One of the main purposes of this book is to provide you with the knowledge of those strategies and approaches that will enhance your

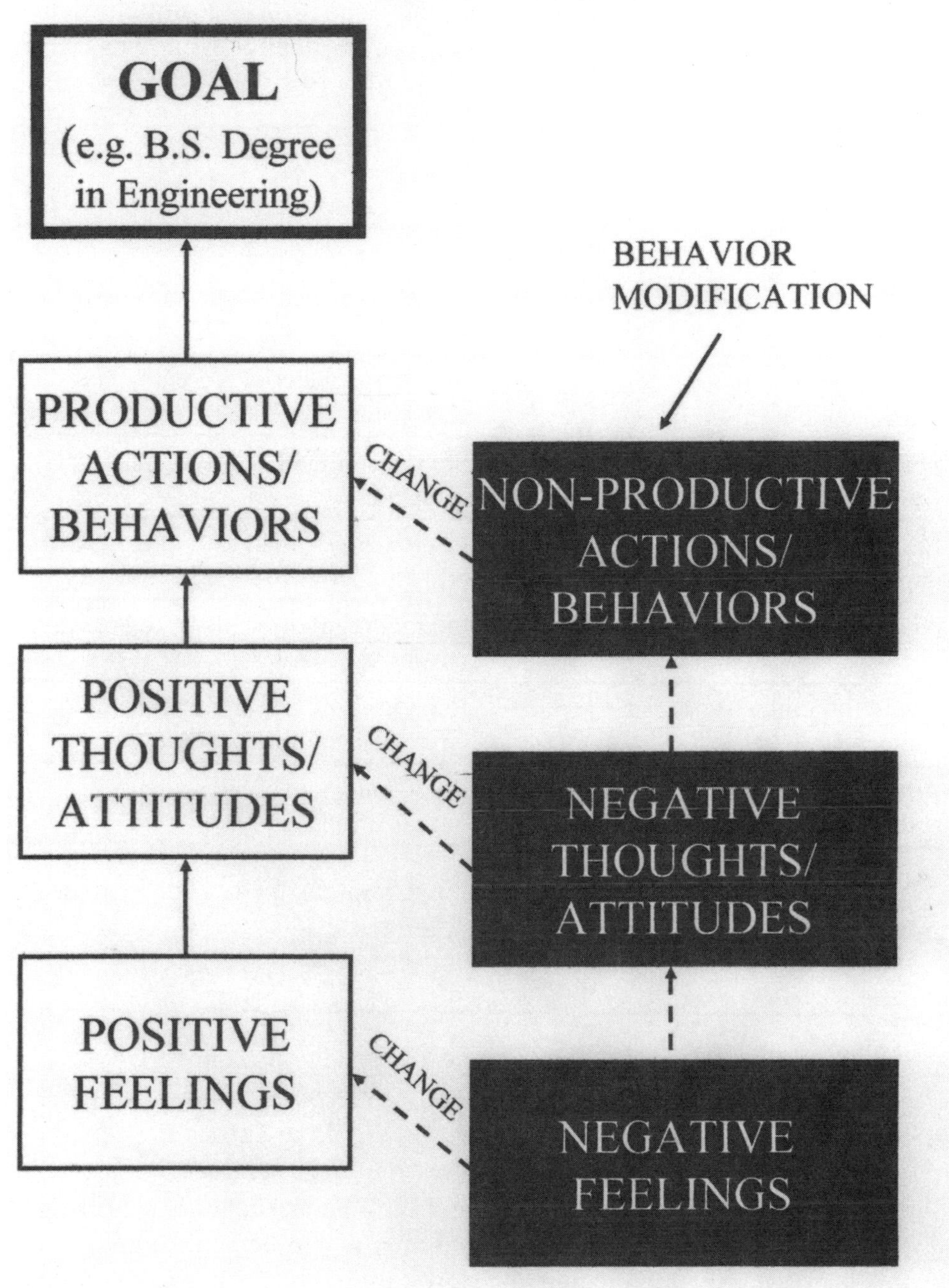
STUDENT SUCCESS MODEL
GOAL
(e.g. B.S. Degree in Engineering)
BEHAVIOR MODIFICATION
PRODUCTIVE ACTIONS/ BEHAVIORS
CHANGE
NON-PRODUCTIVE ACTIONS/ BEHAVIORS
POSITIVE THOUGHTS/ ATTITUDES
CHANGE
NEGATIVE THOUGHTS/ ATTITUDES
POSITIVE FEELINGS
CHANGE
NEGATIVE FEELINGS

effectiveness as an engineering student. Much of that knowledge was presented in Chapters 3, 4, and 5. Hopefully, you have studied those chapters and recognize that implementing many of the strategies and approaches presented there will require you to change your behaviors. Some examples of these changes are presented below.

Non-Productive Action	CHANGE	Productive Action
Neglect studying	⟶	Devote significant time and energy to studying
Delay studying until test is announced	⟶	Master the material presented in each class prior to next class
Study 100% alone	⟶	Study collaboratively with other students
Come to each lecture unprepared	⟶	Review notes, read text, attempt problems prior to each lecture
Avoid professors	⟶	Interact regularly with professors outside the classroom
Spend little time on campus	⟶	Immerse yourself in the academic environment of the institution
Avoid participation in student organizations	⟶	Participate actively in student organizations

Gaining the knowledge presented in Chapters 3, 4, and 5 however, does not guarantee that you put it into practice.

New Knowledge Is No Guarantee

A prime example of how new knowledge does not always produce change is smoking cigarettes. When I was growing up, people simply did not know that smoking caused cancer. Today, no one can deny knowing that it does. The relationship between smoking and lung cancer is an example of a new knowledge base. But many people still smoke. They failed to change because they did not make a commitment to act on the new knowledge.

STEP 2. COMMITMENT—"YOU WANT TO DO IT."

By *commitment,* we mean that you not only know what to do but that:

You want to do it.

Developing that commitment requires you to go through the process of examining each academic success strategy and deciding whether you want to put it into practice. Do I want to schedule my time? Do I want to study from class to class rather than from test to test? Do I want to prepare for each lecture? Do I want to study collaboratively with other students? Do I want to make effective use of my professors? Do I want to spend more time on campus? Do I believe these are strategies and approaches that will enhance my academic success?

A commitment to doing something is really an attitude—or type of thought. The process of making a commitment to the new knowledge base involves making a change in your attitudes. To do this, you must first become conscious of your attitudes, particularly any negative ones that are obstructing your growth. A quote from Deepak Chopra's excellent book *Seven Spiritual Laws of Success* [1] shows the way:

> "Most of us, as a result of conditioning, have repetitious and predictable responses to the stimuli in our environment. Our reactions seem to be automatically triggered by people and circumstances, and we forget that these are still choices that we are making in every moment of our existence. We are simply making these choices unconsciously.
>
> If you step back for a moment and witness the choices you are making as you make those choices, then in just this act of witnessing, you take the whole process from the unconscious realm into the conscious realm. This procedure of conscious choice-making is very empowering."

If negative attitudes are keeping you from changing your behaviors, try the following approach:

(1) Identify key areas in which your attitudes (positive or negative) will have a significant impact on your academic success.

(2) For each area, identify what attitudes you hold.

(3) For each attitude, answer the question, "Is this attitude working for me (positive attitude) or against me (negative attitude)?"

(4) For each negative attitude, answer the question, "Why do I hold this attitude?" (i.e., "Where did it come from?")

(5) For each negative attitude, answer the question, "Can I change this negative attitude to one that will work for me (i.e., a positive attitude)?"

Example of This Process

Let's imagine that a student believes he is failing math because the professor is boring, unprepared, never smiles, and seems aloof. This student has the attitude that "I can't pass a course if I don't like the professor." He sees himself as a "victim" in which passing his math course is viewed as totally in the control of the professor. Once he becomes conscious that this is a negative attitude (one that interferes with his goal of success in school) and he realizes that the attitude can be changed, he can go about changing it to a positive one.

An example of a positive attitude would be: "I can pass a class when I don't like the professor, but it is going to require me to adopt alternate strategies and to put in more work." This positive attitude might lead to behaviors that include sitting in on another instructor's lecture, getting old exams, or seeking help from students who passed the course last semester.

As this example illustrates, changing your attitudes is a necessary precursor to changing your actions.

STEP 3. IMPLEMENTATION—"YOU DO IT!"

The final stage is implementation. By *implementation*, we mean that you not only know what to do and want to do it, but that:

You do it!

This is probably the most difficult step to take. Actual change is hard, no matter how knowledgeable or committed you may be.

There are many reasons why you may fail to change your behaviors that you know you should—behaviors that, in effect, work against you. For example, let's assume that you are not putting sufficient effort into studying and are doing poorly in your classes. How can this be? You want to be successful in school. You want to get all of the rewards that a career in engineering will bring to you. Still, you are not doing what is required for success.

BARRIERS TO IMPLEMENTING PRODUCTIVE ACTIONS

One of the primary reasons you may not be willing to change is that there may be a payoff for you to keep doing what you are doing. You have adopted your current behavior patterns because they satisfy some need or want that you have. Changing to new behaviors will require you to give up old behaviors, ones that you may like very much.

For example, you may go home or to your residence hall as soon as you get out of class. You may do this because you get a great deal of pleasure from the distractions you find there, such as friends, family, TV, music, food in the refrigerator, telephone, Internet. Choosing to stay on campus to study with other students, to seek help from professors, or to use the resources of the library may be less enjoyable to you. You may have to really work on yourself to change your thought from "I enjoy going home" to "If I go home, it's likely that I won't study. I'm going to stay at school until I get my work done."

Or you may have difficulty choosing to do things you don't find easy or enjoyable. Many people do. The impact of choosing to do things we like to do rather than things we don't like to do on our success is perhaps best stated in a remarkable presentation by Albert E.N. Gray to a national meeting of life insurance underwriters in 1940 titled "The Common Denominator of Success" [2] As so well put by Mr. Gray:

> "The common denominator of success—the secret of every [person] who has ever been successful—lies in the fact that they formed the habit of doing things that failures don't like to do."

And Gray goes on to explain why a person would choose to do things they don't like doing:

> "Successful [people] are influenced by the desire for pleasing results. [They] have a purpose strong enough to make them form the habit of doing things they don't like to do in order to accomplish the purpose they want to accomplish. Failures are

influenced by the desire for pleasing methods and are inclined to be satisfied with such results as can be obtained by doing things they like to do."

I hope your commitment to your goal of becoming an engineer is strong enough to influence you to choose productive behaviors, even if you don't like doing some of them. I would also encourage you to read the complete text of Mr. Gray's inspiring presentation. To do so, enter "The Common Denominator of Success" into your favorite Internet search engine and you'll find a number of web sites that feature it.

There are other reasons why you might choose non-productive behaviors. Human psychology is very complicated and doesn't always make sense. You may be "afraid" to study because if you do and still fail, it will reflect on your ability. Or you may be trapped in a "victim" role, preferring to blame your failure on factors or people external to yourself. Perhaps you feel you were forced to go to college by your parents. By not studying you are showing them that you are your own person; that you are not going to do what they want you to do. **Making change requires you to accept responsibility for your actions and to begin to view yourself as the creator of your life.**

REFLECTION

Reflect on Albert E.N. Gray's concept that people who are successful develop the habit of doing things that people who are failures don't like to do. Do you see the wisdom in this idea? What are some things that you don't like to do that need to be done if you are to be successful in your math/science/ engineering studies? Is your commitment to your goal of graduating in engineering strong enough to influence you to develop the habit of doing these things?

6.3 UNDERSTANDING YOURSELF

There is one subject that you don't study in school—a subject that is the key to your happiness and quality of life. That subject is YOU! The most exciting adventure you will ever embark on is the journey of self-discovery and self-awareness.

Understanding yourself is an essential aspect of becoming a productive and happy person. There are other benefits as well. As you grow in your understanding of yourself, you will grow in your ability to understand other people.

But nobody said it would be easy. As Benjamin Franklin noted:

> "There are three things extremely hard:
> steel, a diamond, and to know one's self."

Understanding yourself is a lifelong process. Human beings are very complex. As a result, there are many different models or frameworks available to describe human behavior and human psychology.

Some of these frameworks are more useful than others. In fact, some are not even valid. I always wonder how I am supposed to believe that one-twelfth of the people in the world (all those born under my zodiac sign) have something in common with me. Others are over-generalizations. For example, I have a friend who read a book on the importance of birth order [3]. What a bore he became! He went around asking people their birth order. If they said they were first born, he would tell them that they were reliable, conscientious, driven to succeed, serious, self-reliant, well-organized, and on and on. This is an example of a framework that tries to put all people into one of three categories (i.e., firstborn, middle child, or last born)—an obvious oversimplification.

In Chapter 3, we presented a framework (or model) for the different ways students prefer to receive and process new knowledge. Hopefully, by understanding your preferences, you have grown and developed in your ability to better design your learning process.

In this section, we will present two additional frameworks that can be useful to you as an engineering student. The first is **Maslow's Hierarchy of Needs** [4]. Maslow's Hierarchy will give you an understanding of those needs that must be met if you are to be motivated to succeed in your studies.

One of these needs is the need to feel good about yourself. Because your ***self-esteem*** is a very important factor in your productivity and in your happiness, we will address it in some detail.

Finally, we will discuss the **Myers-Briggs Type Indicator (MBTI)** as a model for characterizing peoples' different personality types, and we will note how engineering students on average differ in personality types from the general population.

MASLOW'S HIERARCHY OF NEEDS

Motivation is an inner drive or impulse that causes you to act in a certain way. Maslow clarified the relationship between motivation and unmet needs. In his theory of motivation, which has become widely accepted, Maslow put forth a hierarchy of needs shown below.

MASLOW'S HIERARCHY OF NEEDS

According to Maslow's theory, *needs* must be satisfied from the bottom up. If a lower-level need exists, you will become highly motivated to satisfy that need. When lower needs are satisfied, higher-level needs become important, and you become motivated to satisfy those needs.

Needs Are Not Wants

> It is important to distinguish between needs and wants. ***Needs*** are things that you must have, things that are essential. ***Wants*** are things that you desire. For example, you may want to have a car, but having one may not be essential. Don't let unnecessary wants distract you from academic success.

At the lowest level are **your physiological needs** for food, water, air, and shelter. Hopefully, you are satisfying these needs. If not, it is unlikely that you will be able to focus on your academic work.

At the second level are **your safety needs**, including the need for security and for freedom from fear of physical and psychological threats. Again, I hope that these needs are satisfied for you. If you are afraid of a bully on your hall or a former boyfriend, it is doubtful that you will be able to concentrate adequately on your studies.

At the third level are **your social needs**, such as needing to belong, to be accepted, and to receive affection and support from others. These social needs are generally met by family or friends. If you left home to go away to school, you may be experiencing a period in which your social needs are not being met. It is important, therefore, for you to develop new friends and relationships at school. Otherwise, unmet social needs can interfere with your studies. Fortunately, many of your classmates are also looking to satisfy these same needs.

At the fourth level are **your needs for esteem**, including self-respect, achievement, and recognition. You need to feel good about yourself and to feel as though you have the respect and appreciation of others. (We will address the important topic of self-esteem in the next section.) Gaining appreciation from others, including your professors and other students, will be related not only to your academic success but to how you treat other people.

At the fifth and highest level is **your need for self-actualization**. *Self-actualization* is full development of your abilities and ambitions. It is the need you have to reach your highest potential, or put in simple terms, "to do your best." This is the need that causes you to want to excel on a test, to do your best in a game of tennis, to learn, to grow, and to develop. Perhaps best put by Maslow:

> "Even if all these needs are satisfied, we may still often expect that a new discontent and restlessness will soon develop, unless the individual is doing what he or she, individually, is fitted for. Musicians must make music, artists must paint, poets must write if they are to ultimately be at peace with themselves. What humans can be, they must be. They must be true to their own nature. This need we may call ***self-actualization***."

Obviously, to be a successful student, you must be able to pursue your need for self-actualization. This means that you must first satisfy your physiological, safety, social, and esteem needs.

Satisfying Your Need for Self-Esteem

As indicated by Maslow's Hierarchy of Needs, self-esteem is a fundamental human need. We cannot be indifferent to the way we feel about ourselves. Self-esteem is a critically important factor to virtually every aspect of our life. It influences what we choose to do, how we treat others, and whether we are happy or not.

Many problems faced by our society such as drug and alcohol abuse, crime and violence, poverty and welfare abuse, teenage pregnancy, the disintegration of the family, and the high dropout rate among high school students are directly related to the low self-esteem of many of our citizens. Several years ago, the California Legislature established a Task Force to Promote Self-Esteem to make recommendations on what the State can do to enhance the self-esteem of its citizens [5]. The Task Force defined *self-esteem* as:

> "Appreciating my own worth and importance and having the character to be accountable for myself and to act responsibly toward others."

According to Nathaniel Branden [6], self-esteem is made up of two interrelated components:

Self-efficacy - your sense of competence

Self-respect - your sense of personal worth

To be self-efficacious is to feel capable of producing a desired result. Self-efficacy is related to your confidence in the functioning of your mind and in your ability to think, understand, learn, and make decisions.

Self-respect comes from feeling positive about your right to live and to be happy, from feeling that you are worthy of the rewards of your actions, and from feeling that you deserve the respect of others.

It is important to have <u>both</u> self-efficacy and self-respect. If you feel competent but not worthy, you may accomplish a great deal, but you will lack the capacity to enjoy it. You may feel that you must continually prove your worth through achievement. Overachievers and "workaholics" are generally striving to meet their need for self-respect by feeling competent and productive.

There is a strong correlation between our self-esteem and our behaviors. According to Branden, healthy self-esteem correlates with:

Rationality	Realism
Intuitiveness	Creativity
Independence	Flexibility
Cooperativeness	Willingness to admit mistakes
Benevolence	Ability to manage change

Poor self-esteem correlates with:

Irrationality	Blindness to reality
Rigidity	Fear of the new and unfamiliar
Rebelliousness	Inappropriate conformity
Defensiveness	Overcontrolling behavior
Fear of others	Hostility toward others

It is no surprise that research has found that high self-esteem is one of the best predictors of personal happiness [6]. The value of self-esteem is not merely that it allows you to feel better, but also that a healthy self-esteem will be a key factor in your productivity and success. According to Branden:

> "**High self-esteem** seeks the challenge and stimulation of worthwhile and demanding goals. Reaching such goals nurtures self-esteem. **Low self-esteem** seeks the safety of the familiar and the undemanding. Confining oneself to the familiar and the undemanding serves to weaken self-esteem."

How can you enhance your self-esteem? The continuous feedback loop between your actions and your self-esteem described by Branden points the direction. The level of your self-esteem influences how you act. Conversely, how you act influences the level of your self-esteem.

Recall our discussion of behavior modification in Section 6.2, which is based on the premise that you **choose** your actions. You can choose productive actions or you can choose non-productive actions. You have less control over your thoughts, but you can catch negative thoughts and work at changing them to positive thoughts. Behavior modification also asserts that if you choose productive actions in support of a personal goal and you work at changing your negative thoughts to positive thoughts in support of those actions, in time your feelings will change. You will feel better about yourself and your life. **Your self-esteem will improve**.

> ***Your college years provide a unique opportunity for you to enhance your self-esteem by building both your self-efficacy and your self-respect.***

Your engineering education will develop your problem-solving skills, your technical knowledge, and your ability to work with others. All of this will increase your confidence in your ability to face life's challenges and to achieve whatever goals you set for yourself. Through this process, you will build your self-efficacy.

You will have many opportunities to build your self-respect and your feeling of personal worth as well. Academic success will bring positive feedback from your professors and from your fellow students. More tangible rewards such as scholarships, internships in industry, and admission to graduate school can be yours. You can be president of an engineering honor society, be the team leader of your institution's entry in a national engineering student competition, be paid to work on a professor's research project, or co-author a paper that is presented at an international conference. These accomplishments will be respected by others and will enhance your sense of self-worth.

Success in engineering study will enhance both your feeling of competence and your self-respect. These together will build a healthy self-esteem. But it is up to you! You can let the negative feelings associated with low self-esteem produce negative thoughts that lead you to non-productive actions and failure. Or you can choose productive actions and positive thoughts that will lead you to success and to feeling good about yourself and your life.

REFLECTION

Consider the ten items listed in this section that correlate with healthy self-esteem. How many of the items would you use to describe yourself? Consider the ten items listed in this section that correlate with poor self-esteem. How many of the items would you use to describe yourself? What can you do to change yourself with regard to any items that correlate with poor self esteem?

MYERS-BRIGGS TYPE INDICATOR (MBTI)

Individuals are different. We each have preferences for how we interact with the world around us, how we learn, how we make decisions.

Personal Example

I, for example, am an extrovert. I am energized by interactions with other people. I like lots going on around me and like to get tasks accomplished. I am a linear thinker. I can figure out anything that is logical. The hardest assignments I had in high school were to memorize poems. I just couldn't do it. It wasn't logical. Furthermore, I'm not very creative. I would be the worst at painting a picture or decorating an empty room. Finally, I like things to be planned and orderly. I don't like surprises. I'm not very spontaneous. When I go on vacation, I make reservations. My worst fear would be to arrive somewhere and find there were no accommodations.

You may be very different from the way I am. You may have very different preferences. People may tire you out. You may prefer to work alone. You may be very creative and artistic. You may prefer being flexible and spontaneous as opposed to being planned and orderly. You may like to arrive in a European city late at night with no reservations. You may be very intuitive, easily grasping the "big picture" through your imagination or bursts of insight.

The famous Swiss psychologist Carl Jung did the seminal work on psychological types [8]. Jung's work led to the Myers-Briggs Type Indicator (MBTI), which is widely used today [9]. The MBTI characterizes individuals in four areas:

(1) Does the person's interest flow mainly to:	
The outer world of actions, objects, and persons?	E-extrovert
The inner world of concepts and ideas?	I-introvert
(2) Does the person prefer to perceive:	
The immediate, real, practical facts of experience and life?	S-sensing
The possibilities, relationships, and meanings of experiences?	N-intuiting
(3) Does the person prefer to make judgments or decisions:	
Objectively, impersonally, considering causes of events and where decisions may lead?	T-thinking
Subjectively and personally, weighing values of choices and how they matter to others?	F-feeling
(4) Does the person prefer mostly to live:	
In a decisive, planned, and orderly way, aiming to regulate and control events?	J-Judging
In a spontaneous, flexible way, aiming to understand life and adapt to it?	P-Perceiving

The result of this characterization is to place individuals into 16 personality types based on combinations of four pairs of letters (E or I, S or N, T or F, J or P).

I would encourage you to not only find out and learn about your personality type, but also to learn about the other 15 personality types. Learning about other personality types will help you understand how others may differ from you.

There are two ways to find out your personality type. One is to take the MBTI. It is very likely that the test is administered somewhere on your campus—either by the testing office or by the career center.

The second way is to take a similar test that is based on the same research that led to the MBTI—the Keirsey Temperament Sorter II. The handy thing about this test is that it can be taken and scored free of charge on the Internet at:

http://www.advisorteam.org

The Keirsey Temperament Sorter II requires you to choose one of two descriptors for each of 70 items. Based on the results, you will be placed into one of the same 16 categories as the MBTI. These are also described in terms of four temperaments, each having four variants, shown as follows with the corresponding Myers-Briggs personality type.

Temperament	Variant 1	Variant 2	Variant 3	Variant 4
Guardian	Supervisor (ESTJ)	Inspector (ISTJ)	Provider (ESFJ)	Protector (ISFJ)
Artisan	Promoter (ESTP)	Crafter (ISTP)	Performer (ESFP)	Composer (ISFP)
Idealist	Teacher (ENFJ)	Counselor (INFJ)	Champion (ENFP)	Healer (INFP)
Rational	Field Marshall (ENTJ)	Mastermind (INTJ)	Inventor (ENTP)	Architect (INTP)

Another feature of the Keirsey web page is that you can access descriptions of each of the 16 personality types. More detailed descriptions can be found in two excellent books that are available through Amazon.com or other sources: Keirsey's book *Please Understand Me II: Temperament, Character, Intelligence* [10] and *Type Talk at Work (Revised): How the 16 Personality Types Determine Your Success on the Job* [11]. By studying these descriptions, you can learn more about yourself and more about how others may be different from you.

You might be interested in knowing that engineering students tend to differ from the general population as shown below [12]:

	General Population	Engineering Students
Introvert (I)	30%	67%
Intuiting (N)	30%	47%
Thinking (T)	50%	75%
Judging (J)	50%	61%

As you can see, engineering students have a greater tendency than the general population to be introverts rather than extroverts, to prefer to use a logical approach to make decisions, and to prefer to live in a planned, orderly way. Although a higher percentage of engineering students are intuitors than the general population, more than half of engineering students prefer to gain information using their senses. In Myers-Briggs "lingo," the most frequent personality type found among engineering students is ISTJ, followed in order by ESTJ, INTJ, INTP, and ENTJ.

6.4 Understanding Others/Respecting Differences

One of the most important areas in which you can strive for personal growth and development is in respecting people who are different from you. Engineering, now more than ever, is a team-oriented profession. Your success both as an engineering student and as an engineering professional will be closely related to your ability to interact effectively with others. As an engineer, you will be required to work with, manage, and be managed by people differing from you in learning styles and personality types and in gender, ethnicity, and cultural background.

Respecting Differences in Learning Styles and Personality Types

Understanding and respecting the different learning styles presented in Chapter 3 and the various personality types presented in Section 6.3 can aid you in both accepting differences in people and in communicating effectively with people who are different from you. Herrmann, in his excellent book *The Creative Brain,* [13] puts forth some pertinent ideas about such differences:

(1) Differences are not only normal, but also positive and creative.

(2) Appreciating and using these mental differences makes change easier to deal with because it makes us more creative.

(3) As we appreciate the full spectrum of mental gifts—ours and those of others—we can make better choices in our lives, especially in selecting educational and career directions.

(4) If those who manage others will acknowledge and honor personal preferences and give people the chance to match their work with their preferences, they will be able to tap tremendous gains in productivity.

(5) As we learn to value and, above all, affirm one another's unique mental gifts, we can participate in the formation of true community—perhaps our best hope for survival in this strife-torn world.

I'm sure you'll agree that this is an impressive list of the benefits that can come to you by better understanding the differences in people's learning styles and personality types.

Ethnic and Gender Differences

Other critically important areas of difference are ethnic and gender differences. Thirty years ago, 98 percent of engineers and computer scientists in the U.S. were white males. This is no longer the case. The percentage of women and ethnic minorities among engineering and computer science graduates has been increasing steadily. The diversity that will soon be reflected throughout the engineering and computer science professions can be seen by looking at the percentage of the 73,602 2004/05 B.S. degrees awarded in engineering and computer science.

As indicated in the chart below, 41 percent of current engineering and computer science graduates are women, ethnic minorities, and foreign nationals. And the percentage is expected to grow in the future.

B.S. Degrees Awarded in Engineering and Computer Science Ethnic and Gender Groups - 2004/05 [15]

	Number	Percentage
Non-Minority Women	7,386	10.0%
African-Americans	3,576	4.9%
Hispanics*	3,860	5.2%
Native-Americans	345	0.5%
Asian-Americans	9,457	12.8%
Foreign Nationals	5,436	7.4%
TOTAL	**30,060**	**40.8%**

* Excluding Puerto Rico

Unfortunately, prejudice, bigotry, and discrimination continue to be prevalent in our society. We seem to have a compulsive need to build ourselves up by putting others down.

How we treat others is closely related to self-esteem. If we don't feel good about ourselves, it is likely that we won't feel good about others. The converse is well stated in the report of the California Task Force to Promote Self-Esteem [5]:

> "The more we appreciate our own worth and importance, the more we are able to recognize and appreciate the worth and importance of others as well."

We now live in a multi-ethnic, multicultural society. The old *melting pot* idea is no longer operative. As a nation, we are now more like a "mosaic," or a mixture of separate and different peoples, each having its own unique characteristics. Indeed, the concept of a *melting pot* has become offensive. Why should I strive to be the same as you? The quality of my life and the contributions I make are related to my special qualities and uniqueness as an individual. If I come from a different background and experience than you, I can act and think in ways that you cannot.

A Very Personal Story

I grew up in a very racist environment. At the time I completed high school in 1957 in Jacksonville, Florida, the worst forms of institutional racism and legalized segregation were in effect. Certainly with that kind of upbringing, I had a lot of changing to do. Several years at a northern liberal college and I became a zealous supporter of the civil rights movement. I came to abhor the evils of the racist society I grew up in. This is perhaps why I have devoted a significant portion of my professional career to working affirmatively to undo the effects of racism.

I could tell so many stories of steps I took in the process of developing my sensitivity to the negative impact of our racist behaviors. One that comes to mind is the lesson I received about ethnic jokes. I used to have my favorite Polish joke. I thought it was a good one, so I always felt like I won points when I told it. When I was working on my Ph.D. degree, I chose Russian to meet my language requirement. I signed up for a course and was a bit late getting to the first class meeting. When I arrived I noticed that the instructor was agitated and upset with the class. As it turned out, he was of Polish descent and apparently the subject of Polish jokes had come up. At the second class period, he brought in a research paper that documented the negative impact of the Polish joke phenomenon in the U.S. on the self-image of Polish-American children. I promise you that I have never told that Polish joke or any other ethnic joke since that day.

As an engineering student and a future engineering professional, you need to learn to respect and value people from different ethnic backgrounds. This may require you to work through certain prejudices that you have. Just remember, a prejudice is nothing more than a thought or attitude that you have the capability of changing.

As we discussed in Section 6.2, knowledge can produce change. Sensitivities can develop by understanding the injustices around us.

For many years, I visited several universities annually to assist them with their minority engineering programs. I always asked to meet with minority students to get their perspective. It broke my heart when they told me that white students wouldn't form laboratory groups with them, left the seats next to them vacant, or acted surprised when they did well on tests. I hope that you will not engage in these very harmful and unnecessary behaviors.

STEREOTYPING IS UNNECESSARY AND UNFAIR

One of the primary issues related to prejudice is ***stereotyping***.

> A ***stereotype*** is a fixed conception of a person or a group that allows for no individuality.

Engineering students are often labeled as *nerds* who care only about things and have no interest in or skill for dealing with people. The obvious problem with stereotyping a group like this is that the stereotype doesn't apply to all individuals in the group.

Just as I'm sure you would not like to be automatically labeled as a *nerd* merely because you are an engineering student, I hope you will refrain from stereotyping others. The best way to approach people who differ from you in ethnicity or gender is to suspend judgment. Take the view that all things are possible. Resist the urge to draw conclusions about someone you don't even know.

Stereotyping also works against gender parity. Sheila Widnall, Secretary of the Air Force in the Clinton Administration, wrote a powerful paper several years ago [16] outlining many of the obstacles experienced by women who study science and engineering. According to Dr. Widnall:

Studies of objective evaluations of the potential and the accomplishments of women give quite discouraging results. Such studies in which male or female names are applied to resumes, proposals, and papers that are then evaluated by both male and female evaluators consistently show that the potential and accomplishments of women are undervalued by both men and women, relative to the same documents with a male attribution.

I hope that you agree that we have a long way to go in providing equal and fair treatment to all people.

Improving Your Effectiveness in Cross-Cultural Communications

What can you do to improve your effectiveness in working with and communicating with people who differ from you? First and foremost, you should seek out opportunities to interact with people from different ethnic and cultural backgrounds. You can learn a great deal from them and improve your interpersonal communication skills in the process. If you really want to grow in this area, take a course in cross-cultural communications. I think you will find the subject very interesting and you will develop skills that will be useful throughout your life.

If we all practiced the *Silver Rule,* originally credited to Confucius, in our interactions with others, we probably wouldn't need to discuss this issue at all:

What you would not want others to do unto you, do not do unto them.

If we practiced this simple principle, we certainly wouldn't put others down, stereotype others, treat others unfairly, resent others, or make others the butts of our jokes, since we would not like to have these done to us.

As a nation, we have made a great deal of progress in the area of race relations and multiculturalism, but not nearly enough. Professor Robert Cottrol of Rutgers Law School gives us an optimistic vision for the future [17].

> "Perhaps our most important contribution to the twenty-first century will be to demonstrate that people from different races, cultures, and ethnic backgrounds can live side by side; retain their uniqueness; and, yet, over time form a new common culture. That has been the American story. It is a history that has much to tell the world."

I hope YOU will contribute to making this vision a reality!

6.5 Assessment of Your Strengths and Areas for Improvement

If you are committed to personal development, you need to start by assessing your strengths and areas for improvement. We talked about this earlier in the chapter when we introduced the concept of a "personal TQM philosophy," and discussed making value judgments about our actions, thoughts, and feelings based on our goals.

Another basis for assessing your strengths and areas for improvement was presented even earlier—in Chapter 1, where we looked at models for viewing your education.

Assessment Based on Attributes Model

If you recall, the **Attributes Model** indicated the knowledge, skills, and attitudes that you should gain from your engineering education. A personal assessment based on this model would involve evaluating how strong you are with regard to each of the following attributes:

An ability to apply knowledge of mathematics, science, and engineering
An ability to design and conduct experiments, as well as to analyze and interpret data
An ability to design a system, component, or process to meet desired needs
An ability to function on multidisciplinary teams
An ability to identify, formulate, and solve engineering problems
An understanding of professional and ethical responsibilities

An ability to communicate effectively
A broad education necessary to understand the impact of engineering solutions in a global and societal context
A recognition of the need for, and an ability to engage in, life-long learning
A knowledge of contemporary issues
An ability to use the techniques, skills, and modern engineering tools necessary for engineering practice

ASSESSMENT BASED ON EMPLOYMENT MODEL

Similarly, you could do a personal assessment based on the **Employment Model**. This model identifies those factors employers will use in evaluating you when you apply for a job as follows:

Personal qualifications (e.g., enthusiasm, initiative, maturity, poise, appearance, integrity, flexibility, and the ability to work with other people)
Special skills or coursework
Grade point average
Communication skills
Engineering-related work experience
Leadership roles in student organizations

ASSESSMENT BASED ON ASTIN STUDENT INVOLVEMENT MODEL

Or you could do a personal assessment based on Astin's **Student Involvement Model**. This model gives metrics for measuring the quality of your education as reflected by your level of "student involvement."

Time and energy devoted to studying
Time spent on campus
Participation in student organizations
Interaction with faculty members
Interaction with other students

How to Do a Personal Assessment

How would you go about doing a personal assessment? It would simply involve rating yourself (e.g., on a scale of 0 to ten; ten being highest) on each item listed. For example, on a scale of 0 to ten, how would you rate your ability to work on teams? How would you rate your ability to identify, formulate, and solve engineering problems? How would you rate your ability to communicate effectively? How would you rate your grade point average?

For those items that get a high mark, just keep on doing what you are doing. We have a tendency to seek out areas in which we are strong. For example, if we have strong computer skills, we like to spend lots of time on the computer. As a result, our strengths are naturally reinforced. There is generally no need to concentrate in a planned way on them.

Personal Development Plan

What we need is to work on our areas for improvement—the very areas we tend to avoid. For example, if we do not write well, we avoid classes that require writing. If we are shy, we avoid people. Avoidance behavior ensures that we will not improve in areas that need improvement.

You probably can't take on all of your areas for improvement at one time. What is needed is to prioritize them in order of importance and choose several of the most important to work on. For each area chosen, create a *personal development plan.* What are you going to do in the next week? In the next month? In the next year?

As an example, if you are shy and lack good interpersonal communication skills, your plan might include some or all of the following action items:

(1) talk more with people

(2) discuss your problem with a counselor in the counseling center

(3) take a course in interpersonal communications

(4) read a book on self-esteem

(5) join the campus Toastmasters Club

(6) take an acting class

(7) join a student organization

The time you are in college is the time to work on your areas for improvement. You can make mistakes there and the price will be low. If you avoid dealing with those areas in which you need to improve, they will follow you into the engineering work world.

There, the price for failure will be much higher!

6.6 DEVELOPING YOUR COMMUNICATION SKILLS

An important area of your personal development is your communication skills. Although you will receive a certain amount of training through your formal education process, it is very likely to fall short of what you really need. In this section, we will explain the importance of strong communication skills in engineering, and suggest ways for you to supplement your required coursework in this area.

THE IMPORTANCE OF COMMUNICATION SKILLS IN ENGINEERING

Consider the following scenarios:

- As a result of your senior engineering design project, you invent a new, high-speed Internet search engine. You want to patent, produce, and market the program before someone else beats you to it.
- With engineering diploma in hand, you are ready to launch your professional career, and you have narrowed your job search to four local engineering firms.
- As an engineer who designs and builds vehicles using alternate energy sources, you are assigned to a team to study the durability of a new composite graphite cloth for the vehicle's body. After six weeks of testing the material, your team is ready to make its recommendations to management.
- Shortly after you graduate with your BSEE degree, you are hired into a position at a prestigious research lab, where your primary responsibility is to develop processes for manufacturing integrated circuits used in communications satellites. Four months into your work, you are selected to present your findings at an upcoming meeting at NASA.

- As a mechanical engineer, you work for a popular amusement park, where you oversee the planning, design, and construction of new rides. In your current project, you supervise a team of 30 engineers and construction workers. Every two weeks, you must submit a written report to your superiors, detailing the progress of the project.

At stake in each of these scenarios is not so much your technical expertise as your ability to communicate—both orally and in writing. To patent, produce, and market the Internet device would involve extensive interactions with all kinds of people, groups, and organizations—from the U.S. Patent Office to prospective clients. To land a job after graduation will depend largely on how well you comport yourself and communicate with others (from your resume and cover letters to follow-up phone calls, interviews, and possibly short presentations). As a member of the engineering team at the alternate-energy vehicle company, you will take part in either writing a technical report or presenting your findings to management in person. The same applies to your meeting at NASA or your progress reports to superiors at the amusement park.

I could easily add many more scenarios to the ones above or create a lengthy list of the writing and speaking demands that you will typically encounter as an engineer. Here are just a few:

- Letters, memoranda, and e-mail correspondences
- Design specifications
- Requests for proposals ("RFPs")
- Proposals submitted in response to RFPs
- Contracts, patents, and other government documents
- Progress reports (written and oral)
- Daily work logs
- Instruction manuals
- Technical reports
- Formal presentations (often to non-engineering audiences)
- Project and committee meetings
- Short courses and training seminars

- Guest lectures at engineering schools or professional engineering society conferences
- Publications of significant findings in professional engineering journals
- Written and oral performance evaluations of subordinates

This list goes on, but I think you see how expansive a role communication skills play in an engineer's world. As a practicing engineer, you will consistently—almost endlessly—draw on your abilities to write and speak effectively.

THE ENGINEERING "DISCOURSE"

The amount and variety of communications, as illustrated above, should be sufficient to convince you of the importance of these skills in engineering. But there is another factor you need to understand in order to appreciate fully the value of strong writing and speaking skills.

In ***all*** communication transactions, regardless of their importance, length, audience, or medium, your performance will ***always*** and ***inevitably*** reflect back on you. Fair or not, your fellow professionals will judge how skillfully you execute every writing or speaking task. If your performance somehow doesn't "measure up," your potential success as an engineering professional will be undermined—and possibly severely limited.

You see, every profession, engineering included, expects certain knowledge, competencies, and an overall presentation of self in order to be accepted into the profession.

James Paul Gee, professor of linguistics at Boston University, explains these professional expectations in terms of a "Discourse" [18]. Gee defines a "Discourse" as

> ". . . a kind of 'identity kit' that includes specific 'knowledge, costumes, and expectations on how to act, talk, and write so as to [exhibit] a particular role that others in the Discourse will recognize.'"

The catch is that, to become a member of a professional Discourse, like engineering, you must be able to demonstrate mastery of ***all*** of the requirements: "Someone cannot participate in a Discourse unless he or she meets all expectations required in that Discourse. Because Discourses are connected with displays of an identity, failing to display **fully** an identity is tantamount to announcing that you don't really have that identity." In short, "you are either in it or you're not." As we have already seen, communication skills play a major role in the engineering profession. You can be sure, then, that your mastery of these skills will be a pivotal factor in determining your eligibility for membership in the engineering Discourse.

EMPLOYERS WANT MORE

Although the term "Discourse" is unfamiliar to engineering employers and educators, they would certainly understand and agree with the concept behind it. For years, employers have complained about the weak communication skills of the engineering graduates they interview, especially when they find a "star" candidate who qualifies in every way except communications. Not surprisingly, a national survey of over 1,000 engineering employers conducted by the National Society of Professional Engineers revealed that industry's #1 concern was to give engineering students "more instruction in written and oral communications" [19].

Educators, too, have long debated ways to improve their engineering students' communication skills. Pressured both by industry's demands and by ABET's criterion that engineering graduates have "an ability to communicate effectively" (see Chapter 1, Section 1.4), engineering educators everywhere are actively engaged in finding solutions to the problem. Some engineering schools are developing upper-division engineering-specific communication courses for their students. Combined with other required composition and speech courses, the creation of "writing-intensive" courses across the curriculum, and (at some institutions) "writing proficiency exit exams," this increased focus on students' communication skills will hopefully produce better prepared graduates. In most cases, though, the engineering curriculum is already so full that additional requirements are not possible. So the debate continues.

THE ENGINEERING STUDENT AS COMMUNICATOR: A PROFILE

Take a moment to reflect on your own communication skills. How would you rate your writing? How do you feel about speaking before audiences or giving oral presentations? Chances are that, as an engineering major, you view writing and public speaking with little enthusiasm. They are probably your least favorite and perhaps weakest subjects, and the fewer courses you must take in them, the happier you will be. After all, you likely chose engineering because your strengths lie in math and science; your engineering curriculum centers largely on developing these strengths; and a "good" engineer is one who excels in these areas.

Although you now understand the need for strong communication skills, you may lack confidence or feel discouraged about your ability to improve them. Many new engineering students do, especially those whose native language is not English. To make matters worse, you are bound to encounter fellow engineering students whose communication skills clearly outshine your own. How, you wonder, can you ever learn to communicate as well as they do? Your required coursework in communications will help, but will that be enough? Probably not.

The answer is that you ***can*** become an effective communicator. The skills are not all that difficult to master, but they do require two commitments from you:

(1) A positive attitude

(2) A 4-5 year plan to ensure regular practice

DEVELOPING A POSITIVE ATTITUDE

Let's address attitude first. The profile above, highlighting the fear, intimidation, discouragement, and perceived inability that many engineering students feel about their communication skills, is quite understandable. Most people, regardless of profession, dislike writing of any kind and abhor public speaking. It's not surprising. Writing and speaking are like "baring your soul" to others—often strangers. To some extent, you can hide weak writing skills by hiring an editor or "ghost" writer. You also have the opportunity to proofread and revise—again and again—before making a written document public. Speaking before an audience, however, offers no such safety nets. Like a ballet dancer or tight-rope walker, you need only one wrong step to ruin your performance.

While understandable, the fears associated with writing and public speaking are counterproductive—particularly for you, as future engineers. I can't tell you how many times I have seen promising engineering students try to capitalize on their technical knowledge as a way to compensate for inadequate communication skills. Because they feared or disliked writing and speaking, they did as little as possible in college to develop these skills. When, as seniors, they began their career search, they repeatedly lost out to others who had wisely balanced their engineering education with training and practice in communications.

It would be great—wouldn't it?—if we could change our negative attitudes about writing and speaking instantaneously. Just one "ZAP!" from Batman or the wave of a magician's wand and—voila!—we can't wait to write a technical report or make a presentation before a group of executives. But, while we may lack such magical swiftness, we do have the capacity to change negative attitudes into positive ones. In terms of writing and speaking, you need just one little success—an "A" grade on a composition or a compliment from a professor about your contributions in class—to start the process. Make one of these small challenges a goal of yours this semester, and watch how your attitude about writing and speaking begins to change.

Developing a Plan to Improve Your Communication Skills

You've heard the expression that "practice makes perfect." It's true. No matter how adept or advanced you may be in a skill, you need to practice it regularly to maintain it. Just ask any professional athlete.

Strong skills of any kind also require time to grow: years; not days, weeks, or even months. That's why you need to start TODAY to build your communication skills. The longer you delay tending to them, the weaker they will be when you graduate—and the longer it will take you to prove your eligibility to join the engineering Discourse.

What should you do? Lay out a 4-5 year plan that includes as many courses that focus on communications skills as you can work into your program. Build in your required courses first, like freshman composition and speech communications. Then browse through your university catalog for additional communications courses: advanced writing and speech courses, business communications courses, psychology/human relations courses, theater courses, "writing-intensive" courses, and so forth. See which of these courses could be used to meet your general education requirements. You may also have some free electives that you can devote to developing your communication skills. Another option would be to take a few "extra courses" in this important developmental area that don't meet graduation requirements. An ideal goal would be to take one course in this area each term during your four-year program.

Once your plan is complete, put it into action. Approach each class with the same enthusiasm and interest you bring to your engineering courses. Sit in the front of the classroom. Be inquisitive: ask questions. When faced with a writing assignment, do it early so that you can get feedback from your instructor or tutor, and rewrite it based on their suggestions. Do the same for oral assignments. Practice an upcoming presentation before a group of friends or classmates. Even better, videotape yourself and then evaluate your performance. We're always our own toughest critics!

Beyond formal coursework, look for extracurricular opportunities to write and speak. Practice your writing by keeping a journal or corresponding with friends via e-mail. Write a poem or short story. *Write a critique of this book and send it to me!* Above all, **read**—anything and everything. Research in language development has shown that reading is a significant way to improve writing skills. Read the newspaper, magazines, technical journals, and novels. Set goals for your reading, particularly during breaks between school terms. Take the time you would normally watch television and use it for recreational reading.

Develop your oral communication skills with the same vitality and engagement, both in and out of the classroom. A speech class will introduce you to the field of rhetoric and give you practice in the rhetorical "modes of discourse" (e.g., narrating, describing, analyzing, persuading, and arguing). Psychology courses will teach you the principles of human relations, group dynamics, and cross-cultural communications. A theater course will give you instruction and practice in effective delivery.

Extracurricular activities are also plentiful and beneficial. Getting involved in engineering student organizations or school athletic programs will go far in building your interpersonal and teamwork skills. Running for student offices and holding positions in the student body government will strengthen your public speaking and presentation skills. Outside of school, scrutinize speakers you hear on television and radio. Study the techniques of famous speakers, like Dr. Martin Luther King, Jr. and John F. Kennedy, or *any* individual whose speaking skills you admire. Take stock of their strengths and weaknesses, and try to incorporate some of their strategies into your own speaking repertoire.

CONCLUSION

Whatever avenues appeal most to you to develop your communication skills, what's most important is that you **DO SOMETHING** and **START TODAY**. A 4-5 year plan developed ***now*** will help keep you on track. If you sincerely commit to this plan, I guarantee that your most defeatist attitude will change for the better. Watch your self-confidence and poise grow. Note how your fellow engineering students start to admire you and seek your "secret" to success. And don't be surprised when prospective employers start vying for your attention.

6.7 Mental and Physical Wellness

To be productive and happy, it is important that you take care of yourself personally. With the rigors and demands of being a student, it is easy to ignore your emotional and physical well-being. But that is a big mistake! Tending to your personal needs is a must.

Many people are not aware of the connection between our physical and emotional health. The fact is they are strongly interrelated: our physical well-being greatly affects our emotional state—and vice versa. For example, one of the best remedies for emotional stress is vigorous physical exercise. And I'm sure you've noticed that when you are mentally "up," you tend to feel good physically, whereas if you're emotionally down, you often feel physically fatigued or even get sick.

Tips for Good Health

Since each of us is so unique and our emotional and physical states so complicated, this section is only meant to offer you a few ideas. Most obviously and most importantly, to expect a high level of mental and physical health, you must:

- **Eat nutritionally**
- **Engage in regular aerobic exercise**
- **Get adequate sleep**
- **Avoid drugs**

Eat Nutritionally. What you eat significantly affects your physical and mental state. A proper diet consists of fresh fruits and vegetables, lean meat in moderation, and whole grain products. Avoid processed foods, fatty foods, and sugar. Not only will you feel better now but you'll reduce your chances of heart attack, cancer, and other diseases later.

Engage in Regular Aerobic Exercise. Regular aerobic exercise in which you get your heart rate above 130 beats per minute for more than 20 minutes at least three times a week is essential to good physical condition. If you're not already engaged in some form of exercise, you should consider taking up jogging, brisk walking, swimming, biking, rowing, aerobic dancing, spinning, or any vigorous activity that will improve your physical fitness and that you do regularly.

Get Adequate Sleep. Different people require different amounts of sleep and the amount needed may change as you grow older. Only you can determine how much sleep you need. Just remember that your work efficiency will decrease if you are getting either less or more sleep than you need.

Avoid Drugs. Drugs are abundant in our society. Some, such as caffeine, alcohol, and nicotine, are legal; others, such as marijuana and cocaine, are illegal. Regardless of their legality, all can be harmful and my advice to you is simple: avoid them. Not only do drugs detract from your physical and mental health, they also can greatly interfere with your ability to study.

Balancing Work and Play

To ensure a healthy mental state, you need to **strike a balance between immediate and future gratification**. By seeking too much immediate gratification, and therefore not getting your work done, you are likely to feel guilty. You'll probably then worry about the fact that you are not studying, putting yourself in a mental state in which you cannot study.

On the other hand if you work too much, too long, or too hard, you begin to feel deprived. Feelings of deprivation and resentment can begin to sabotage your commitment. You may begin to doubt whether the sacrifice is worth it.

What you need to find is a proper balance between work and play. One approach is to tie work and play together through a system of work and rewards. Rewards can be small things, like taking a break, going for a walk, watching your favorite TV show, taking an hour for recreational reading. Or they could be larger things, such as going to a party, buying yourself some new clothes, or going away for the weekend with a friend. The point is that, rather than take the view that the work you are doing will not have a payoff until far into the future, you provide yourself with frequent and immediate rewards for your hard work.

Managing Stress

Interestingly, the term *stress* was borrowed from engineering by Dr. Hans Selye [20], an early pioneer in the area of stress management. Selye defined stress as "the response of the body to any demand made upon it to adapt, whether that demand produces pleasure or pain."

Stress can be externally imposed or internally imposed. Certainly, the announcement by three of your professors that each has scheduled a midterm exam on the same day can create stress for you. Causes of internally imposed stress include unmet expectations, high personal standards, irrational ideas, and unrealistic demands you place on yourself.

Stress can be either positive or negative. *Eustress* is a positive form of stress that motivates individuals to attain higher levels of performance. The "butterflies" a football player experiences before the big game can produce inspired play. *Distress* is the negative form of stress. It can distract you from being the best that you can be. It can debilitate and be devastating to one's physical and mental health.

Some common causes of stress include worry, frustration, anxiety, and depression. Frustration is our response to being prevented from gratifying certain impulses or desires. For example, it would be frustrating if you were unable to enroll in a mathematics course you needed to take. Worry and anxiety are closely related. Both are your response to a perceived threat. Anxiety is a somewhat stronger emotion. You become worried if your roommate fails to come home for several days. You become anxious when you develop pains in your chest. Depression is an extreme form of worry and is an emotional condition characterized by feelings of hopelessness and inadequacy.

Each of these emotions is a potential source of stress (stressor). Stressors do not affect everyone in the same way. What would cause stress for one person may not even bother another. Your reaction to stressors is undoubtedly related to your self-esteem. If you feel competent and worthy, you are in a good position to handle stress. Your reaction to stressors is also tied to how "in control" or "out of control" you feel about your life. For individuals who lack self-esteem and do not feel in control of their lives, stress can produce anger, depression, and physical illness.

Whether *eustress* or *distress*, Selye demonstrated that stress produces the "fight or flight" response in our body. This is an instinctive physical reaction to threat, either physical or psychological, which we inherited from our ancient ancestors. Under stress, blood is diverted to the brain and muscles for clearer thinking and quicker reflexes, the heart rate accelerates, the blood pressure rises, the respiration rate increases, and the pupils of the eyes dilate. For engineering students, this response is inappropriate, since you will neither need to "fight" nor "take flight." The stressors we're talking about are mental or emotional ones, so the concepts behind "fight or flight" are, at best, metaphoric. Still, it is interesting to see the interactions between perceived stressors and one's immediate response to them.

To bolster your effectiveness as an engineering student, it is important that you learn to cope with and manage stress. Kaplan [21] has drawn up 11 quick stress dischargers you can use to deal with short-term stress:

(1)	Have a good cry
(2)	Create something manually
(3)	Talk it out
(4)	Have some fun
(5)	Take a walk
(6)	Get a massage
(7)	Take a hot bath
(8)	Breathe more slowly
(9)	Learn to relax
(10)	Turn to your friends
(11)	Groan deeply

It is important that you use practical strategies such as these to prevent the "burn-out" that can come from unchecked or unresolved stress. In the long term, good nutrition, regular exercise, relaxation, and good time management will help to keep your stress level low.

If all else fails, it is important that you seek counseling or medical treatment. Extreme stress can lead to severe physical incapacitation—often to the point of prolonged diseases or even death.

6.8 Motivating Yourself

We will conclude this chapter on student development with several inspirational messages. If you accept the premise of behavior modification that you hold the reins for becoming the best you can be (i.e., achieving self-actualization), perhaps the messages of others who have succeeded will motivate you to do the same.

A Very Personal Story

Sometimes I wish I could "bottle and sell" the feeling I have about the value of my education. I owe almost everything of quality in my life to my education. I have had so many unique and rewarding experiences, so many challenges, so many opportunities. I've been paid well to do work I enjoy. I have been able to travel, write, speak, teach, and influence others. I have gotten to know many interesting people. I have had options and choices and control over where I live and what I do. I can hardly imagine going through life without an education.

Often I wonder why I have been so fortunate. How is it that I went so far with my education? Where did my motivation come from? Actually, it is as clear as day. My mother gave it to me. And she did it in such a dramatic way that, as I look back, there was no possibility I would not go to college and no possibility I would not succeed. If I did fail, I would have let down my mother's #1 dream.

How did she motivate me so strongly? I'll tell you. When I was two years old, my father died. He had just changed jobs and so my mother received only a very modest pension of $15/month.

As a single parent, my mother could have really used that money, but every single month she put $3.75 with that $15 and purchased a $25 United States Savings Bond and put it away for my education. Every month she took me to the bank to put the savings bond in her safe deposit box and told me that that money was for my college education.

Your parents or guardians may have sent you similar messages about the importance of education. If they did, you are indeed fortunate. You are very likely to succeed.

Not all students have parents or guardians who sent them positive messages about the value of a college education. I have known students, in fact, whose parents were opposed to their going to college, preferring instead that they work to help support the family.

However, regardless of any messages you received as a youngster about the value of a college education (and the consequent success it will bring you), you are now an adult. You can think for yourself. You can define your own reasons for wanting to get your education. You can motivate yourself.

We already addressed the idea of motivating yourself through an awareness of the rewards and opportunities an engineering career offers. In Chapter 2, we discussed my "top ten" list of what it will mean to the quality of your life if you are successful in completing your engineering degree. The following sections present additional perspectives regarding the value of your education.

"No Deposit, No Return"

When I was Director of the Minority Engineering Program at California State University, Northridge, we had a motivational button based on the logo shown below.

The clenched fist holding the slide rule (a symbol of engineering in years past) represents the power that comes to an individual through a technical education. The motto "No Deposit, No Return" reminds us, however, that this power does not come without a "deposit." Getting a college education is not easy, and majoring in engineering is even more difficult. Without a doubt, the "deposit" you must make as an engineering major is a rigorous, demanding one. But I guarantee that you will be paid back for the investment many, many times over.

JESSE JACKSON'S "EXCEL" MESSAGE

The Reverend Jesse Jackson has spent much of his professional career motivating young students. His *Push for Excellence* program had a significant impact on African-American youth in Chicago and across the nation.

Reverend Jackson's basic message is that you should strive to "excel" in everything you do. But what does it mean to excel? Very simply, it means

Do your personal best!

If your "personal best" produces a "C" grade in a course, then you have excelled in that course. But how many of us do our best? Do you?

There are reasons why you should strive to excel—to do your best. We have already discussed some:

- It will enhance your self-esteem.
- It will be good for society.
- You'll have a challenging and rewarding career.

Can you think of others?

POWER OF POSITIVE THINKING

I like to make an analogy between jogging and going to college. People take up jogging because they perceive certain benefits. They expect to live longer, feel better, breathe easier, and lose weight. Initially they may dislike the experience of jogging, suffering through it solely for the end result. Eventually, however, most joggers learn to enjoy the experience. They come to enjoy the physical elements of jogging—the rhythmic cadences of moving and breathing, the harmony between body and mind—and they find that long periods of jogging can lead to particularly unique experiences: the so-called "runners' high," a heightened sensitivity to the world around them, an ability to think creatively and imaginatively.

As an engineering student, you can liken yourself to a jogger. At first you may resent or dislike the college experience but you persevere because of the future benefits you anticipate—career opportunities, money, social status, security. But you will eventually come to appreciate your time in school, not only for the benefits it promises, but for the experience itself.

If, like the novice jogger, you find that you dislike school, you are not focusing on the positive aspects of being a college student. You need to recognize that you have created an attitude that may have nothing to do with reality. In fact, you probably are in the best situation of your life and just not aware of it. Surely you have heard people say that their college years were the best years of their lives. Why do you suppose they say this?

If you do have a negative attitude toward school, now is the time to change it. For it's more than likely that you are neither performing at your peak effectiveness nor enjoying what should be a most exciting, rewarding time in your life. Learn to focus on the positive aspects of being a college student. Some of the most significant of these aspects are:

GROWTH PERIOD. As a college student, you are in an unusually heavy growth period. One indication of this is the way in which you are outgrowing your friends from high school who are not going to college. Probably never in your life will you be in such an intense period of learning and personal growth as when you are in college.

EXPOSURE TO PEOPLE. College puts you in an extremely people-oriented environment. Never again will you be with so many people of the same age and interests. You'll find that many of the friends you make during your college years will be important and helpful to you throughout your life.

MANAGER OF YOUR TIME. As a college student you are working for yourself. You have no boss, no one to tell you what to do. Except for your class time, you are pretty much free to manage your time and your affairs.

STARTS AND STOPS. School starts and stops, somewhat like the running of a race. When the race starts you put out a great deal of effort, maybe more than you would like to, but you do so because you can see that it will end. When it does end, you then have an extended period of time for rest and rejuvenation—a break you will not have once you start working as an engineer.

When you learn to appreciate these and other unique aspects of being a college student, you will see an improvement in your academic performance. Remember that:

Positive attitudes bring positive results!

Negative attitudes bring negative results!

Reflection

Reflect on the following quote from Winston Churchill: "The pessimist sees difficulty in every opportunity. The optimist sees the opportunity in every difficulty." Which statement best describes you? Do you tend to see difficulties in every opportunity? Or do you tend to see opportunities in every difficulty?

Summary

This chapter focused on your personal growth and development. Because change is such a critical factor in personal growth and development, we first discussed the psychology of change. We particularly noted how people's receptiveness to change has grown dramatically in recent years—a credit owed largely to U.S. business and industry's embracing the philosophy of "continuous improvement" and its derivative program, Total Quality Management (TQM).

We then proposed a personal adaptation of TQM, which we called "student development." The particulars of each individual's student development plan will certainly vary, but all are based on the following premises:

(1) Your goals set the context and direction of your personal development plan.

(2) You monitor the progress of your personal development by a measurement system that analyzes and critiques your actions, thoughts, and feelings.

(3) You not only implement the plan and track your progress; you also continuously revise and update the plan to ensure your "continuous improvement."

Next, we discussed mechanisms for human behavioral change and presented one based on behavior modification theory—a three-step process that is the most practical and accessible for students. Those three steps entail (1) knowledge, (2) commitment, and (3) implementation. As part of this presentation, we also discussed barriers to change that you are likely to encounter as you move through the three steps.

A necessary backdrop to behavioral and attitudinal changes is self-understanding, so we discussed various models of human behavior to help you understand yourself better.

Maslow's Hierarchy of Needs showed you what basic human needs must be met before you can undertake a plan of self-improvement and growth. Along with this discussion of basic needs, we pointed out the role self-esteem plays in the process of personal growth, development, and change.

The Myers-Briggs Type Indicator (and the closely related Keirsey Temperament Sorter II) was presented as another way to understand yourself. This instrument gives insight into personality types.

The topic of understanding others and respecting differences in people was also discussed. An important component of your personal growth is your effectiveness in working with people who differ from you.

Assessment as a tool to identify and work on areas in which you are weak was discussed next. We focused on two particular areas that are often overlooked in personal development: communication skills (i.e., writing and speaking) and physical and mental wellness. In this latter area, we emphasized the importance of managing stress.

We concluded by offering several motivational messages to under gird your commitment to the difficult process of behavioral and attitudinal change. In conjunction with these messages, we highlighted the positive aspects of being a college student. Our overall intention was to stress how change leads to personal growth, how personal growth leads to the accomplishment of goals, and how the accomplishment of goals leads to success—in school, in the engineering profession, and in life in general.

References

1. Chopra, Deepak, *The Seven Spiritual Laws of Success*, Amber-Allen Publishing & New World Library, San Rafael. CA, 1994.

2. Gray, Albert E.N., "The Common Denominator of Success," Presentation to National Association of Life Underwriters, Philadelphia, PA., 1940.
(Available at: http://www.thinkarete.com/wisdom/works/essays/1462)

3. Lehman, K., *The Birth Order Book: Why You Are the Way You Are*, Revell Books, Grand Rapids, MI, 2004.

4. Maslow, A., *Motivation and Personality*, Harper and Row, New York, NY, 1970.

5. "Toward a State of Esteem: The Final Report of the California Task Force to Promote Self-Esteem and Personal and Social Responsibility," California State Department of Education, Sacramento, CA, 1990.

6. Branden, Nathaniel, *The Six Pillars of Self-Esteem*, Bantam Books, New York, NY, 1994.

7. Myers, David.G., *The Pursuit of Happiness: who is happy—and why*, W. Morrow, New York, NY, 1992.

8. Jung, C.G., Adler, G., and Hull, R.F.C., *Psychological Types (Collected Works of C.G. Yung Volume 6)*, Princeton University Press, Princeton, NJ, 1976 (originally published in 1921).

9. Briggs, K.C. and Myers I.B., "Myers Briggs Type Indicator, Form G," Consulting Psychologists Press, Palo Alto, CA, 1977.

10. Keirsey, David, *Please Understand Me II: Temperament, Character, Intelligence,* Prometheus Nemesis Book Company, 1998.

11. Kroeger, O, Thuesen, J. M., and Rutledge, H., *Type Talk at Work (Revised): How the 16 Personality Types Determine Your Success on the Job*, Delta, 2002.

12. Wankat, P.C. and Oreovicz, F.S., *Teaching Engineering*, McGraw-Hill, 1993.

13. Herrmann, Ned, *The Creative Brain*, Ned Herrmann/Brain Books, Lake Lure, NC, 1990.

14. Lumsdaine, Edward and Lumsdaine, Monika, *Creative Problem Solving: Thinking Skills for a Changing World,* Second Edition, p. 102, McGraw-Hill, 1993.

15. Gibbons, Michael, "The Year in Numbers," American Society for Engineering Education, Washington, DC, 2005. http://www.asee.org/publications/profiles/upload/2005ProfileEng.pdf

16. Widnall, Sheila E., "AAAS Presidential Lecture: Voices from the Pipeline," *Science*, Vol. 241, pp. 1740-1745, September, 1988.

17. Cottrol, Robert, "America the Multicultural," *The American Educator,* Winter, 1990.

18. Gee, James Paul, "Meditations on Papers Redefining the Social in Composition Theory," *The Writing Instructor*, p. 177, Summer 1989.

19. "Report on Surveys of Opinions by Engineering Deans and Employers of Engineering Graduates on the First Professional Degree," Sustaining University Program, National Society of Professional Engineers, Publication No. 3059, Alexandria, VA, November, 1992.

20. Selye, H., *Stress without Distress*, J.B. Lippincott, Philadelphia, PA, 1974.

21. Kaplan, Myron S., "Beating the Training—Stress Connection," *Data Training,* March, 1988.

PROBLEMS

1. After doing some research, write a 500- to 750-word essay on "Total Quality Management." How does TQM compare to the management approaches you have observed in the university you attend, organizations you belong to, or companies you have worked for?

2. For the next week, write down any negative thoughts you are able to identify. Describe at least one non-productive action that is likely to be the result of each negative thought. Also write down a positive thought that you could substitute for the negative thought. Describe at least one productive action that is likely to result from each positive thought.

3. Scrutinize your behavior in the past week and then write down five non-productive actions you have engaged in. Why did you choose each of them?

4. When we receive a stimulus (e.g., see food), we often act (e.g., eat the food) and then think about our action (e.g., "I shouldn't have eaten the food"). How can you change this order (stimulus ⇒ action ⇒

thought) in a way that it might lead you to choose productive actions more frequently?

5. Add ten examples of productive actions (actions that will enhance your academic success) to the list of seven used as examples in Section 6.2.

6. Do you think "behavior modification" can work for you? Why? Do you think "counseling/therapy" would benefit you? Why?

7. Convert the following negative thoughts to positive thoughts by finding a higher context from which to view the situation that led to the negative thought:

   a. I wish I were taller.
   b. I'm homesick.
   c. I don't have any friends.
   d. My chemistry lectures are boring.
   e. I don't know if I like engineering.
   f. I wish I could find a better roommate.
   g. I don't have time to exercise regularly.

8. Examine the following productive behaviors:

   a. Study collaboratively with other students.
   b. Devote significant time and energy to studying.
   c. Prepare for each lecture.
   d. Study from class to class rather than from test to test.
   e. Make effective use of professors outside of the classroom.
   f. Practice good time management principles.
   g. Immerse yourself in the academic environment of the institution.
   h. Actively participate in student organizations.

   Do you:

   (1) Have adequate knowledge about each behavior?
   (2) Have you made a commitment to the behavior? If not, why not?
   (3) Are you implementing the behavior? If not, why not?

9. Consider Maslow's Hierarchy of Needs. How well are your needs being met at each level? Think up one or two ways to better meet your needs at each level. Do them.

10. Research the term *self-actualization*. What does it mean? How strong is your need for self-actualization?

11. Consider the ten items listed in Section 6.3 that correlate with healthy self-esteem. Write down a brief definition of each item. How many of the items would you use to describe yourself?

12. Consider the ten items listed in Section 6.3 that correlate with poor self-esteem. Write down a brief definition of each item. How many of the items would you use to describe yourself?

13. Based on the results of Problems 11 and 12, would you say that you have a healthy self-esteem or a poor self-esteem? Explain. What can you do to improve your self-esteem?

14. Write a paragraph describing yourself in terms of the four personality indicators that are measured through the Myers-Briggs Type Indicator (MBTI). From this analysis, what MBTI indicator (e.g., ENFP) do you think best describes you?

15. List five ways you benefit from knowing your MBTI personality type.

16. Find out if you can take the MBTI on your campus (at the testing office, counseling center, etc.). If you can, take the test and determine your MBTI personality type. Compare the results with Problem 14.

17. Take the Keirsey Temperament Sorter II on line at:

    http://www.advisorteam.org

    What temperament type are you? Find the description of that temperament type on the Keirsey web site. Does it describe you? Print out the description and ask someone who knows you well whether it describes you.

18. Go to the Keirsey web site (address in Problem #17) and read the description of the Guardian Inspector (ISTJ) and the Idealist Healer (INFP). Do you think an ISTJ would make a good engineer? What about an INFP?

19. List five reasons why you should strive to improve your effectiveness in working with and communicating with people who are different from you.

20. Conduct a personal assessment based on the "Attributes Model" by rating yourself on a scale of 0 to ten (ten being highest) on each of the 11 attributes listed in Section 6.5. Identify the five areas in which you rate the lowest. Pick the three of those that you feel are the most important. Develop a personal development plan to improve in each of these areas.

21. Evaluate yourself, on a scale of 0 to ten (ten being highest), with regard to the following personal qualifications:

    a. Enthusiasm
    b. Initiative
    c. Maturity
    d. Poise
    e. Integrity
    f. Flexibility
    g. Ability to work with other people

    What can you do to improve yourself in the areas in which you have given yourself a low evaluation?

22. Assess the quality of your education as measured by Astin's "Student Involvement Model." Give yourself a rating of 0 to ten (ten being highest) for each of the five areas listed in Section 6.5. Develop a plan for improving in each of the areas you feel you are weak. Implement the plan.

23. List ten types of documents that an engineer might have to write. Which of these do you feel qualified to write at this time?

24. Write a proposal seeking funding (e.g., from parents, private foundation, scholarship committee, etc.) to support your education. Explain how much money you need, why it's needed, and how giving you the money will ultimately benefit the *funding source.*

25. Develop a personal development plan for improving your writing skills over the next three to five years. Implement the plan.

26. Pick one of the personal development books listed at the end of Chapter 6 (Reference 1, 6, 7, 10, 11, or 13). Read the entire book, or a specific part of it, and write a critique of what you read.

27. Develop a personal development plan for improving your oral communication skills over the next three to five years in each of the following areas:

    a. Interpersonal communications
    b. Group communications
    c. Formal presentations

    How do your plans differ for each of these areas?

28. You are hired to conduct a telephone campaign to request contributions from alumni of your engineering school. Write an

opening statement and a series of follow-up statements that you feel would persuade an alum to make a contribution.

29. One strategy for improving your vocabulary is to write down words you don't know from things you read, look them up in the dictionary, and try to add them to your vocabulary. Carry this strategy out for the following words which appeared in Chapter 6:

    paradigm
    conduce
    discourse
    efficacy
    attribution
    stereotype
    cathectic

30. Conduct a personal assessment based on the four keys to good health listed in Section 6.7:

    a. Eat nutritionally
    b. Engage in regular aerobic exercise
    c. Get adequate sleep
    d. Avoid drugs

    Develop a personal development plan for any areas in which you rate yourself low.

31. For the next week, schedule your study time in blocks. Following each block of study time, schedule time to do something you enjoy as a reward for doing your work. At the end of the week, evaluate how this plan worked for you.

32. Write a paragraph describing the messages your parents sent you about the value of a college education. Do you agree with these messages?

33. In addition to the four positive aspects of being a college student discussed in Section 6.8, list five others. Do you believe the adage that "Positive attitudes bring positive results; negative attitudes bring negative results"? If you do, why? If you don't, why not?

CHAPTER 7
Broadening Your Education

We are not just our behavior.
We are the person managing our behavior.

Kenneth Blanchard

INTRODUCTION

How do you view your education? A narrow view would be that you get an education by passing a prescribed set of courses. A quality education is much more than that. This chapter will introduce you to ways you can broaden your education and in doing so be significantly better prepared for a successful career.

First, we will discuss the value of participation in student organizations and extracurricular life. The skills you can develop through these activities could be as important to your success as those you gain through your formal academic work.

Next, we will describe opportunities for you to gain practical engineering experience through participation in student design competitions, technical paper contests, Design Clinics, and research projects.

Then we will discuss strategies and approaches for seeking pre-professional employment, including summer jobs, part-time jobs, and cooperative education work experiences. Through such pre-professional employment, you can gain valuable practical experience, better define your career goals, and earn money to support the cost of your education.

Finally, we will discuss opportunities for you to put something back into the educational system of which you are a part. Opportunities for service can range from visiting high schools to recruit students to providing feedback to faculty so they can improve the quality of their teaching.

7.1 Participation in Student Organizations

In Chapter 1, we presented several models for viewing your education. Both Astin's *Student Involvement Model* and the *Employment Model* indicated that you should participate actively in student organizations and extracurricular campus life. The reasons were slightly different for each model, but very much related. The *Involvement Model* indicated that the quality of your education will be enhanced through participation in student organizations. The *Employment Model* identified experience in campus activities, particularly participation and leadership in extracurricular life, as an important factor used by employers in evaluating candidates for employment. Employers place a high value on such participation because it signals your leadership and organizational skills.

If you do not participate actively in student organizations, I expect you have you own reasons. Perhaps you have never thought of becoming involved. Or perhaps you are not aware of the benefits of such participation. On the other hand, maybe you have considered becoming involved but decided that you don't have enough time. Or maybe you are shy, and therefore reluctant to join a group of people you don't know.

I hope I can persuade you to let go of these or other reasons you might have. Participation in student organizations can contribute significantly to the quality of your education. Through such participation, you can:

- Meet your social needs
- Develop your leadership and organizational skills
- Engage in professional development activities
- Receive academic support
- Participate in service activities

And, once again, these benefits are highly valued by employers.

> Imagine yourself, as you near graduation, interviewing for a position with a local company. You can bet one of the questions you will be asked by the interviewer is, "Can you give me any examples of your involvement in student organizations, particularly those in which you took on leadership roles?" How do you think it will be viewed when you answer, *"Not really. I was too busy studying."*

A word of caution, however. Be selective about your involvement in such activities, since the opportunities to participate are so numerous that you could wind up neglecting your studies.

Your university could have literally hundreds of student organizations. These include recreational organizations, service organizations, social fraternities and sororities, ethnic- and gender-based organizations, and academic and professional organizations.

ENGINEERING STUDENT ORGANIZATIONS

Of the many different student organizations, the ones that are the most accessible to you and have the greatest potential for benefit are the academic and professional student organizations that operate within your engineering college. Most of these engineering student organizations fall into one of three categories:

(1) Student chapters of discipline-specific national engineering societies

(2) Engineering honor societies

(3) Student chapters of national ethnic- and gender-based engineering organizations

STUDENT CHAPTERS OF DISCIPLINE-SPECIFIC ENGINEERING SOCIETIES. It is very likely that your engineering college has a student chapter corresponding to your engineering discipline. For example, if you are an electrical or computer engineering major, you could join the Institute of Electrical and Electronics Engineers (IEEE) student chapter. As a mechanical engineering major, you would want to become involved in the American Society of Mechanical Engineers (ASME) student chapter, and so forth.

There could be several different student organizations corresponding to a particular discipline, each representing a specialization within that discipline. For example, in addition to the ASME student chapter, the mechanical engineering department at your institution could also be home for a Society of Manufacturing Engineers (SME) student chapter, a Society of Automotive Engineers (SAE) student chapter, a Society for the Advancement of Material and Process Engineering (SAMPE) student chapter, and an American Society of Heating, Refrigerating and Air-Conditioning Engineers (ASHRAE) student chapter. Similarly, a civil engineering department could have, in addition to an American Society of Civil Engineers (ASCE) student chapter, student chapters of the Structural Engineers Association (SEA) and the Institute of Transportation Engineers (ITE).

ENGINEERING HONOR SOCIETIES. Each engineering discipline also has an honor society. Examples of honor societies for specific disciplines are:

Civil Engineering - *Chi Epsilon*

Electrical and Computer Engineering - *Eta Kappa Nu*

Mechanical Engineering - *Pi Tau Sigma*

Industrial Engineering - *Alpha Pi Mu*

Chemical Engineering - *Omega Chi Epsilon*

Aerospace Engineering - *Sigma Gamma Tau*

Geological and Mining Engineering - *Sigma Gamma Epsilon*

Materials Science and Engineering - *Alpha Sigma Mu*

In addition to honor societies for each discipline, there is an honor society covering all engineering disciplines: *Tau Beta Pi*. *Tau Beta Pi* is the engineering counterpart of *Phi Beta Kappa*, the honor society for liberal arts students.

You cannot choose to join honor societies; instead you must be invited. These invitations are extended to junior or senior students who have achieved an academic record that places them in the top ten or 20 percent of students in their major.

I encourage you to set a personal goal of gaining membership in *Tau Beta Pi*. Striving to be in the top 20 percent of your peers is a lofty yet achievable goal, and one that is well worth shooting for. Membership in *Tau Beta Pi* is a very prestigious honor. There is almost nothing you can put on your resume that will impress employers more than membership in

Tau Beta Pi. To learn more about *Tau Beta Pi*, visit the society's web page at:

http://www.tbp.org

ETHNIC- AND GENDER-BASED STUDENT ORGANIZATIONS. Your engineering college may also have one or more ethnic-based engineering student organizations. The most common of these organizations are:

National Society of Black Engineers (NSBE)

Society of Hispanic Professional Engineers (SHPE)

Society of Mexican-American Engineers and Scientists (MAES)

American-Indian Science and Engineering Society (AISES)

The purpose of these organizations is to increase the representation of these ethnic groups in the engineering profession. However, membership is not restricted, and all who are committed to the purpose of the organization are welcome.

Your campus may also have a student chapter of:

Society of Women Engineers (SWE)

The purpose of SWE is to increase the representation of women in the engineering profession. Again, membership is open to all students.

Each of the above five ethic- and gender-based student organizations operate under the auspices of a national organization. You can learn more about these organizations by visiting their web sites:

National Society of Black Engineers - http://www.nsbe.org

Society of Hispanic Professional Engineers - http://www.shpe.org

Society of Mexican-American Engineers and Scientists - http://www.maes-natl.org

American-Indian Society of Engineers and Scientists - http://www.aises.org

Society of Women Engineers - http://www.swe.org

These web sites also contain information about how to start a new student chapter of the organization. If a student chapter does not currently exist on your campus, you may be just the person to get one started.

ENGINEERING STUDENT COUNCIL. All of the engineering student organizations on your campus may be organized into an engineering student council. The purpose of such an *umbrella* organization is to coordinate activities sponsored jointly by the student organizations such as industry career days or events held during National Engineers Week.

BENEFITS OF PARTICIPATION IN STUDENT ORGANIZATIONS.

When you join an organization that is a student chapter of a national engineering professional society, either discipline-based or ethnic- or gender-based, (e.g., IEEE, ASME, ASCE, SHPE, SWE), by paying your dues you become a student member of the national organization, and you will benefit from student activity programs conducted by the professional society. You will receive society publications and, in some cases, student magazines. You will be eligible to attend local, regional, and national meetings and conferences of the society. You will be eligible to compete for various awards, scholarships, and fellowships. You also will be eligible to use any career guidance or job placement services offered by the national organization.

But your greatest benefits will come from participating in the activities of your campus's student chapter. These benefits fall into five major categories:

(1)	**Social interaction**
(2)	**Personal development**
(3)	**Professional development**
(4)	**Academic development**
(5)	**Service to the college and the community**

Let's explore each of these briefly.

SOCIAL INTERACTION. Participation in engineering student organizations can help you develop relationships with students who have similar backgrounds, interests, and academic and career goals as you. Close association with such students can enhance your academic success through the sharing of information and group study. And relationships you develop with fellow engineering students can continue long after your college days are over.

Student organizations promote this social interaction through social functions such as mixers, parties, picnics, and athletic competitions. Fund-raising activities such as car washes, raffles, jog-a-thons, and banquets also facilitate social interaction among members. Many organizations have a student lounge or study center that can greatly enhance the social environment for members.

Personal Development. Through participation in engineering student organizations, you can develop leadership, organizational, and interpersonal skills so important to your success as an engineering professional. You will learn from your involvement in student organizations that it is a significant challenge to get a group of people to agree on a direction and move efficiently in that direction. As you learn to do this, you will acquire important skills in communicating, persuading, listening, cooperating, delegating, reporting, managing, and scheduling.

An engineering student organization can assist members in developing leadership and organizational skills by conducting leadership workshops or retreats, and by sponsoring speakers and seminars on organizational management. The greatest lessons, however, result from opportunities to practice leadership skills. Such opportunities can be provided to the maximum number of members by putting a committee structure in place to accomplish the various objectives appropriate to an organization (e.g., Membership Committee, Social Committee, Professional Development Committee, Academic Support Committee, High School Outreach Committee, etc.).

Professional Development. Participation in engineering student organizations can enhance your understanding of the engineering profession and the engineering work world. Much of the material presented in Chapter 2 can be brought to life through professional development activities conducted by student organizations.

Student organizations can sponsor speakers and field trips to industry. They can organize career day programs in which industry representatives meet with students to discuss employment opportunities. And they can sponsor workshops in important career development areas such as resume writing, interviewing skills, and job search strategies.

Academic Development. Participation in engineering student organizations can enhance your academic performance through direct academic support activities.

Student organizations can sponsor mentor programs in which upper-division (junior and senior) students assist lower-division (freshman and sophomore) students. Organizations can arrange for volunteer tutors, organize review sessions and study groups, and maintain files of lab reports, exams, and homework. Some student organizations have their own study space to promote group study and sharing of information among members. By establishing group challenge goals, such as an overall average GPA target for members, student organizations can motivate members to excel academically.

SERVICE TO THE COLLEGE AND THE COMMUNITY. Participation in engineering student organizations can provide you a vehicle for service to the engineering college and the surrounding community.

Student organizations can sponsor visits to high schools to recruit students into engineering, raise money to establish a scholarship, sponsor activities during National Engineers Week, organize a "Meet the Dean" event, or perform other service projects that benefit the engineering college.

PARTICIPATION IN OTHER EXTRACURRICULAR ACTIVITIES

Beyond participation in engineering student organizations, there are other extracurricular activities that can contribute to your personal development. Examples of these are writing for your campus newspaper, joining a debate club, or participating in musical or dramatic productions.

And don't forget student government—another excellent opportunity for personal growth and development. Eventually you may want to run for one of the many elected offices, but many leadership positions are filled through appointments. Visit your Associated Students office, and ask how you can become involved. ***Who knows? Maybe in a few years you'll be running for student body president!***

7.2 PARTICIPATION IN ENGINEERING PROJECTS

The quality of your education can be significantly enhanced through participation in engineering projects, contests, and competitions. Four of such opportunities will be discussed here:

(1) Student design competitions
(2) Technical paper contests
(3) Design clinics
(4) Undergraduate research

STUDENT DESIGN COMPETITIONS

In recent years, the number of engineering student design competitions has grown steadily. Some of these are paper studies, but most involve the design and fabrication of an engineering device, often followed by competition against entries from other universities. Many of these competitions are sponsored by the professional engineering societies listed at the end of Chapter 2. Most of the competitions involve teams of engineering students rather than individual student participation. Most have prizes, like trophies or cash awards.

Participation in one of these design competitions will give you practical "real-world" engineering experience. You will learn to work on a complex project, subject to strict deadlines, and requiring a high degree of cooperation and coordination. You will experience the type of design tradeoffs and difficult decisions that are characteristic of engineering projects. A significant investment of time will be required, but the rewards will be well worth the effort.

Your engineering college may participate in one or more of these competitions already. Check with the appropriate engineering department. If you have an interest in an event that your college does not participate in, you may just be the catalyst to persuade them to do so.

The following is a representative list of these engineering student design competitions, including the name of the contest, a brief description of it, and its sponsor. To obtain more information, either go to web site listed or enter the name of the competition into a search engine such as Google.

Competition	Description	Sponsor
Reduced Gravity Student Flight Opportunities	Propose, design, fabricate, fly, and assess a reduced-gravity experiment	**NASA Johnson Space Center** http://microgravityuniversity.jsc.nasa.gov
National Student Design Competition (Individual and Team)	Solve a problem that typifies a real, working, chemical engineering design situation	**American Institute of Chemical Engineers** http://www.aiche.org/Students/Awards/index.aspx

ChemE-Car Competition	Design and construct a chemically powered vehicle to carry a specified cargo a specified distance	**American Institute of Chemical Engineers** http://www.aiche.org/Students/Awards/ChemeCar.aspx
Human-Powered Helicopter Competition	$20,000 (one time only) prize for controlled flight (hover for 1 minute; reach 3 meter altitude; stay in 10 meter square)	**American Helicopter Society** http://www.vtol.org/awards/hph.html
Helicopter Design Competition	Design a rotorcraft which meets specific requirements	**American Helicopter Society** in conjunction with NASA and rotating sponsors: Sikorsky, Boeing, and Bell Helicopter http://www.vtol.org/awards/sdcomp.html
1/4-Scale Tractor Design Competition	Design and build a 1/4-scale tractor	**American Society of Agricultural and Biological Engineers** http://www.asabe.org/students/tractor/asaecomp.html
AGCO Student Design Competition	Design an engineering project useful to agriculture	**American Society of Agricultural and Biological Engineers** http://www.asabe.org/awards/competitions/National.html
Challenge X: Crossover to Sustainable Mobility	Re-engineer a GM vehicle to minimize energy consumption and emissions while maintaining utility and performance	**U.S. Department of Energy/General Motors** http://www.challengex.org/about/index.html
Solar Splash	Design, build, and race solar-powered boat	**Solar Splash HQ** http://www.solarsplash.com
Solar Decathlon	Design, build, and operate the most attractive and energy-efficient solar-powered home	**U.S. Department of Energy** http://www.eere.energy.gov/solar_decathlon

ASME Student Design Competitions	Design, construct and operate a prototype meeting the requirements of a problems statement	**American Society of Mechanical Engineers** http://www.asme.org/Events/Contests/DesignContest/Student_Design_Competition.cfm
World HPV Championships	Design, build, and race a streamlined bicycle	**International Human Powered Vehicle Association** http://www.ihpva.org
The Great Moonbuggy Race	Design and build a vehicle that can carry two students over a half-mile simulated lunar terrain course	**NASA Marshall Space Flight Center/Northrop Grumman** http://moonbuggy.msfc.nasa.gov
Concrete Canoe	Regional and national competition to design, build, and race a canoe constructed from concrete	**American Society of Civil Engineers** http://www.asce.org/inside/nccc.cfm
AIAA Design Competitions	Several annual design competitions in engine design, aircraft design, and space satellite design	**American Institute of Aeronautics and Astronautics** http://www.aiaa.org/content.cfm?pageid=210
ASCE/AISC National Student Steel Bridge Competition	Design a steel bridge that meets given specifications and build a 1:10 scale model	**ASCE and American Institute of Steel Construction (AISC)** http://www.asce.org/students/stud_comp.cfm (Click on "National Student Steel Bridge Competition")
MicroMouse	Design and build a micro-processor-controlled robot capable of negotiating a specified maze	**IEEE** Student Chapters organize events. Info at: http://www.micromouseinfo.com/index.html
International Submarine Races	Design and build human powered submarine to compete in open ocean or model basin	**Foundation for Undersea Research and Education** http://www.isrsubrace.org

Aero Design	Conceive, design, fabricate, and test a radio-controlled aircraft that can take off and land while carrying the maximum cargo	**Society of Automotive Engineers** http://students.sae.org/competitions/aerodesign
Super Mileage	Design and build a vehicle powered by a 3.5-HP Briggs & Stratton engine fuel economy record	**Society of Automotive Engineers** http://students.sae.org/competitions/supermileage
Clean Snowmobile Challenge	Re-engineer an existing snowmobile for improved emissions and noise while maintaining or improving performance	**Society of Automotive Engineers** http://students.sae.org/competitions/snow
Formula SAE	Conceive, design, fabricate, and compete with small formula-style racing cars	**Society of Automotive Engineers** http://students.sae.org/competitions/formulaseries
Baja SAE	Design, build, and race off-road vehicle powered by 10-horsepower Briggs and Stratton engine	**Society of Automotive Engineers** http://students.sae.org/competitions/bajasae
Collegiate Inventors Competition	Submit original and inventive new ideas, processes, or technologies	**National Inventors Hall of Fame Foundation** http://www.invent.org/Collegiate
Rube Goldberg Contest	Design and build a machine that uses the most complex process to complete a simple task	**Purdue University** http://www.purdue.edu/UNS/rube/rube.index.html
North American Solar Challenge	Design, build, and race a solar-powered electric vehicle	**U.S. Dept of Energy/ Natural Resources Canada** http://americansolarchallenge.org/index.html
World Solar Challenge	Solar-electric vehicle race across Australian outback every two years (Next race fall, 2007)	**Panasonic/South Australia Government** http://www.wsc.org.au

REFLECTION

Read the brief descriptions of each of the 27 engineering student design competitions listed above. Take a few seconds right now and imagine yourself involved in the design and fabrication of a human-powered helicopter, a solar-powered race boat, or a microprocessor driven robot that could negotiate a maze. Think through what would be involved in taking one of these projects from start to successful completion. These projects are excellent examples of the exciting challenges engineering offers

TECHNICAL PAPER CONTESTS

Many of the professional engineering societies listed at the end of Chapter 2 sponsor technical paper contests. The contests are conducted annually and generally start with a regional competition. In most cases, regional winners progress to a national or international competition. Cash prizes are given to the top-place finishers in both the regional and national competitions. Following are some of the societies that sponsor such contests:

American Institute of Chemical Engineers
American Society of Agricultural and Biological Engineers
American Society of Mechanical Engineers
American Society of Civil Engineers
Institute of Electrical and Electronics Engineers
Institute of Industrial Engineers
Society of Petroleum Engineers
The Minerals, Metals, and Materials Society

For more information on society-sponsored student paper contests, visit the society web page listed at the end of Chapter 2, inquire at the engineering department office, or go see the faculty advisor of your engineering student chapter.

DESIGN CLINICS

The "Design Clinic" concept was pioneered at Harvey Mudd College [1] and has been adopted by a growing number of engineering colleges across the country. Through participation in a Design Clinic, you can work as part of a team of undergraduate students on a problem provided by industry. Design Clinic problems could involve an engineering design problem, software development, performance testing of a product, or a theoretical study.

The Design Clinic work conducted by a student team is supervised by a faculty advisor. Funding is generally provided to the university by the company sponsoring the project. In some cases, students working on a Design Clinic may be paid. In other cases, they may receive academic credit.

Design Clinics are a true "win-win" situation. Companies win by having important problems solved at a modest cost. Companies also are given the opportunity to observe the work of some of the best engineering students. As a student participating in a Design Clinic, you win by having the opportunity to work on a practical industry problem under the supervision of a faculty member. You also benefit from contact with practicing engineers and from getting a "foot in the door" of a company you may want to work for when you graduate.

Undergraduate Research

You can also broaden your education by working for engineering professors on their research projects. Research projects differ from Design Clinics in that research generally involves creating and organizing new knowledge and disseminating that knowledge through publications in technical journals and presentations at scholarly meetings.

Most likely, you only see engineering faculty as teachers. You may not realize that they are also expected to conduct research. The amount of research expected of engineering professors varies from one university to the next.

Most of this research work is supported by external funding. Professors submit proposals to outside agencies requesting money to cover the costs of conducting the proposed research. One of the primary uses of the funding is to hire students to do the work. Although most of the students working on research projects are graduate students pursuing either their M.S. or Ph.D. degree, opportunities also exist for undergraduate students.

I encourage you to seek out opportunities to work on research projects during your period of undergraduate study. Openings may be listed at your university career center, but more likely you will have to speak to individual professors or ask about funded projects at various engineering department offices.

The benefits of an undergraduate research experience can be significant. You will earn money to support the costs of your education. You will have the opportunity to work closely with an engineering professor. Since other students will probably be working on the project, you will learn how to work as part of a team. An undergraduate research experience also gives you a chance to "try out" research to see if graduate school is for you. Depending on the nature of the project, you will develop your skills in specific areas such as laboratory work, computing, or engineering analysis. It is possible that you will be listed as a co-author on papers resulting from the research, and you may even have the opportunity to present the results of your work at a student research conference.

7.3 PRE-PROFESSIONAL EMPLOYMENT

The *Employment Model* presented in Chapter 1 indicated six factors that employers consider in selecting individuals for employment. One of these was engineering-related work experience.

A company considering you for employment when you graduate would like to see that you have had previous work experience, preferably engineering-related. Engineering-related work experience not only demonstrates interest, initiative, and commitment on your part; it also provides you with references—people you have worked for who can vouch for your abilities. Prospective employers also feel that the experience you have gained will reduce the time it takes for you to become productive in their company.

BENEFITS OF PRE-PROFESSIONAL EMPLOYMENT

Pre-professional employment can benefit you in many other ways. Most obvious is that you will **earn money** to support the cost of your education. In addition, the process of seeking pre-professional employment can be viewed as a **rehearsal for the search** you will eventually conduct for a permanent job. You will **develop important skills** related to preparing yourself for a job search, identifying potential employers, and presenting yourself in the best light to those employers.

Pre-professional employment will enhance your **professional development** as well. You will **gain exposure to engineering practice** that will assist you in selecting your major course of study. You will gain a **better understanding** of the various engineering job functions. You will have an opportunity to **apply your knowledge, skills, and abilities**. On your return to school, you will **better understand** how your academic coursework relates to the engineering work world. All of this should increase your motivation to succeed in engineering study.

TYPES OF PRE-PROFESSIONAL EMPLOYMENT

Pre-professional employment can take the form of:

Summer jobs

Part-time jobs

Cooperative education ("co-op") experiences

Each of these is briefly discussed in the following sections.

SUMMER JOBS. Most engineering employers hire engineering students during the summer. Many employers have a formal summer job program in which they bring in a specific number of students each summer.

An engineering-related summer job will not only provide you with the many benefits discussed in the previous section; it will provide you a welcome break from the grind of the academic year. Summer can be a time for rejuvenation. After a meaningful summer work experience, you are likely to return to school re-energized with renewed commitment.

I suggest that you set a personal goal of working in an engineering-related summer job for one or more of the summers during the period of your undergraduate study. You need to realize, though, that student demand for summer jobs outpaces the supply. So you need to adopt the positive, assertive attitude that if anyone is going to get a summer job, it's going to be you. Approaches for conducting a job search are presented in subsequent sections of this chapter.

PART-TIME JOBS. You may also want to work in an engineering-related job on a part-time basis during the academic year. The availability of engineering-related jobs will depend on the location of your university. If your university is located in a major urban area, opportunities may be abundant. In contrast, if it is located in a small town, there may be no engineering employers within commuting distance.

Often students who do well in a summer job are invited to continue working on a part-time basis during the academic year. Although the employer may benefit by having you continue to work, this may not be the best situation for you. It can be flattering to be invited to continue working during the academic year, and the money may be tempting. Just make sure that you make a wise decision—one that takes your overall academic and career goals into account.

There are some tradeoffs to consider when choosing between working in a non-professional job on campus and an engineering-related job off campus. The on-campus job will take less of your time since you will not have to commute. And it will probably be easier to fit in a few hours here and there. On the other hand, you will get more relevant experience from an engineering-related job, and the pay will probably be better.

If you do decide to work on campus, try to find a job that will complement your academic work. Working as a tutor, peer counselor, teaching assistant or grader, undergraduate research assistant, or engineering lab assistant are examples of such positions.

One final thing to remember: full-time engineering study is a full-time commitment. You can probably work up to ten hours per week and take a full course load. If you work more than ten hours per week, you should consider reducing your course load. Recall the guidelines from Chapter 1:

Hours worked	Max course load
10 hrs/wk	full load
20 hrs/wk	12 units
40 hrs/wk	8 units

Keep in mind that these are only guidelines. There are students who are able to work full-time and take a full load of courses. You will have to experiment with what works for you given your individual ability, background, energy level, and willingness to make personal sacrifices.

COOPERATIVE EDUCATION. The federal government defines *cooperative education* as [2]:

> "a program of study at an institution of higher education under which regular students undertake academic study for specified periods of time alternating with work experience in government, industry, business . . . ."

The work periods can range from part-time work while engaging in part-time study (*parallel* co-op) to the more traditional "six-months-on, six-months-off" (*traditional* co-op) arrangement. The opportunity for parallel co-op is generally limited to universities located in areas having many nearby engineering employers.

Cooperative education provides students with some distinct benefits. Among them are:

- Practical experience in industry
- Money to support college expenses
- A "foot in the door" in terms of seeking permanent employment upon graduation

Traditional co-op experiences will provide you with all the benefits described above, but because of the more lengthy period of full-time employment, the experience gained is generally more meaningful than summer or part-time jobs. More challenging assignments can be given as progressively more experience is gained over the six-month period. The benefits of co-op are even more pronounced when the student participates in a second or third co-op experience at the same company.

The "down sides" of co-op are minimal. Participation in one or more traditional cooperative education work experiences will delay your graduation by up to one year. Also, some students have difficulty adjusting to their return to the university after a co-op experience. Six months of earning a good salary and having your nights and weekends free can become habit-forming.

At some universities, co-op is a mandatory part of the engineering program. At most universities, however, participation in a co-op work experience is something that the student may elect. Often, students receive academic credit for the co-op assignment, but this can vary from institution to institution.

The degree of assistance that universities provide to students seeking placement in co-op positions also varies. Universities that have a mandatory co-op program will generally have a well-staffed engineering co-op office that identifies co-op positions and matches students with those positions. At the other extreme, students may be virtually on their own to find co-op positions with minimal help from their career center.

HOW DO YOU MEASURE UP?

Regardless of the form of pre-professional employment you seek, your competitive position will be based on three main factors:

(1) Your year in school
(2) Your academic performance
(3) Your personal qualifications

As a freshman or sophomore, you will have more difficulty finding employment because companies generally prefer juniors and seniors—students closer to graduation. Junior and senior students bring a stronger technical background to their work. And the company values the opportunity to take a look at a student who will soon be a candidate for permanent employment.

But you can make up for your freshman or sophomore status by being strong in items #2 and #3. If you are a top student academically, companies will be interested in developing an early relationship with you. The competition for top engineering students is keen. Companies are well aware that hiring you after your freshman or sophomore year will give them the "inside track." The question they will ask themselves is, "Are you worth the longer wait?"

Your personal qualifications will be a major factor in your success in landing a pre-professional employment position. The "bottom line" question prospective employers will ask themselves is, ***"Will we enjoy having this student in our organization?"*** The answer will be based on an overall evaluation of your enthusiasm, initiative, communication skills, and ability to work with others. An employer will not be disappointed if you fail to solve their most pressing technical problem. But they will be very disappointed if you bring a negative, uncooperative, or unfriendly attitude to your work.

Regardless of how you measure up against the three factors discussed above, your chances of landing an engineering-related job while you are a student will depend to a great extent on how you go about your job search. Effort and approach were discussed in Chapter 1 as keys to your academic success. Similarly, effort and approach will be key factors in your success in landing a summer job, part-time job, or co-op position. Conducting a job search not only takes considerable time and effort; it also requires that you put into practice certain strategies and approaches.

A job search can be divided into the following steps:

(1) Preparing yourself for the job search
(2) Identifying opportunities
(3) Applying for positions
(4) Following up on interviews

Each of these steps in the job search process is described in the following sections.

PREPARING YOURSELF FOR A JOB SEARCH

Aside from getting the best grades you possibly can and developing yourself personally using the principles presented in Chapter 6, there are specific things you need to do to prepare yourself for a job search. You need to develop a resume, learn how to write cover letters, and hone your interviewing skills.

PREPARING A RESUME. A resume is a written document that lists your work experience, skills, and educational background. A resume is your main vehicle for presenting yourself to a potential employer. The central question to ask in preparing your resume is, "If you were an employer, would you want to read this resume?" Employers generally prefer well-written, one-page chronological resumes. Visual impact and appearance are extremely important. Content should include:

Identifying data (name, mailing address, telephone number, and e-mail address)

Employment objective

Education to date

Work experience

Specialized skills

Activities and affiliations

Honors and awards

Assistance in developing your resume should be available through your career center. You can also find extensive resources regarding resume preparation including formats and templates on the Internet. One such source is the CollegeGrad.com web site:

http://www.collegegrad.com/resumes/quickstart/template.shtml

There you will find sample resumes for various engineering disciplines that will guide you in preparing your resume. There are also a number of excellent books on resume writing [3, 4].

REFLECTION

Locate a template for your resume by either going to the CollegeGrad.com web site indicated above or by putting "resume template" or "resume format" into your favorite search engine such as Google. Make up your resume. Make a commitment to having your resume "ready-to-go" at all times during your period of engineering study. You never know when it will come in handy.

PREPARING A COVER LETTER. Whether you contact prospective employers by e-mail, regular mail, or in person, you should always include a cover letter with your resume. And you should create a customized cover letter for each resume you send out. According to an article in the *National Business Employment Weekly* [5], if you want your cover letter "to score a direct hit in your quest for interviews":

- Write to a specific person in the firm, using name and title. This should be the person who makes the hiring decision or for whom you'd work, if hired.
- In your opening paragraph, write something that demonstrates your knowledge of the organization and shows that your letter isn't a form letter.
- Communicate something about yourself that relates to the employer's needs and discusses what you can contribute.
- Ask for a meeting (don't call it an interview). In your closing, be sure to state that you would like to meet the person and will call in a few days to schedule a time.
- Limit your letter to one page, preferably printed on personalized stationery.

One additional "must": as you prepare your cover letter, pay careful attention to your organization of ideas, grammar, spelling, and the overall appearance of the letter. Many employers use cover letters to evaluate candidates' writing skills and professionalism.

DEVELOPING YOUR INTERVIEWING SKILLS. The final area in which you need to prepare yourself is in the area of interviewing skills. Think of an interview like the *final examination* in a course. You wouldn't consider taking a final exam without extensive preparation.

If want to gain some feedback on whether or not you are prepared to perform well in an interview, seek out a friend or fellow student and have them ask you the following questions:

How would you describe yourself?
What are your long-term career goals?
Why did you choose engineering as your major?
How would you describe your ideal job?
Why should I hire you?
What was your favorite course?
Have you ever had a professor you didn't like? Explain.
What is your grade point average?
Have you taken on any leadership roles in student organizations?
Can you give any examples of working effectively with a team of students?
What are your greatest strengths?
What are your major shortcomings?
How would your skills meet our needs?
What have you accomplished that you are the proudest of?
What would you like to know about us?

I expect that the above exercise will convince you of the need to put significant effort into preparing yourself for interviews. You can find lots more questions by entering "interview questions" into your favorite search engine such as Google.

A Personal Story

When I applied for the position of Dean of Engineering, I spent an enormous amount of time preparing for the interview. I called a number of deans of engineering and asked them to tell me about the important issues facing engineering education.

I then put together a list of questions I expected to be asked, and prepared written answers for each one. I practiced answering the questions on anyone I could get to ask them and sought their critique of my answers. I think I knew more about the dean's job on the day of my interview than I did after serving in the position for many years!

In addition to practicing questions and answers, there are other ways to prepare for an interview. Learn as much as you can about the company, the job you are seeking, and the person who will be interviewing you. This task is so much easier than it used to be because of the wealth of information readily available to you on each company's web site. And if the web site doesn't answer all your questions, research the company at your career center or call the company directly. Also, be sure to develop a list of questions to ask the interviewer. Being inquisitive and asking good questions are sure ways to impress a person.

Aside from providing you information about a company, your campus career center can assist you in other critical ways. Most career centers offer "interviewing workshops" and "mock interviews." In mock interviews, you are queried by a staff member assuming the role of a corporate recruiter, who then gives you valuable feedback.

Another way for you to gain insight into how well you interview is to videotape yourself responding to interview questions. Videotaping is a powerful tool that you should try to use.

Finally, there are many excellent references that can help you develop your interviewing skills [6,7].

IDENTIFYING EMPLOYMENT OPPORTUNITIES

There are many avenues you can take to identify pre-professional employment opportunities. Your career center is a good place to start.

One of the career center's main functions is to arrange on-campus interviews. Although most of the interview opportunities will be for students seeking permanent employment, some may include interviews for summer or part-time jobs. Even if your career center doesn't provide these opportunities, it will have a list of companies that conduct on-campus interviews for engineering graduates—an excellent source of leads for you to pursue on your own. Your career center probably also maintains a bulletin board on which it posts pre-professional job listings.

Another strategy for identifying pre-professional job opportunities is to attend any job fairs or career day programs held on your campus. Try to establish personal relationships with the industry representatives there. Be friendly and sell yourself—maybe wrangle an invitation to visit their facility.

NETWORKING. Networking is "the process of exchanging information, contacts, and experience for professional purposes." In plain language, it means talking to friends, fellow students, seniors about to graduate, professors, or anyone else who might have information about job availabilities.

Other candidates for networking include practicing engineers who come to your campus to give a talk or engineering professionals you encounter at meetings of engineering societies. And don't forget alumni of your engineering program. Your college or university alumni association or individual professors can help you identify successful alumni. But anyone, from neighbors and relatives to your doctor or people you know through church or other community affiliations, may be able to open a door for you.

Studies of successful job searches indicate that networking is one of the best ways to find a job.

View everyone you know or meet as a possible lead to a job!

Remember, people enjoy helping others. If you ask people for advice, they will gladly offer it. One warning, however. People do not like to be responsible for others. So don't make others feel that getting you a job is their responsibility.

One last point about networking. Don't think of it as a one-way street. Just as others can be a resource for you, you can be a resource for others. As you progress through the process of preparing and searching for jobs, you will gain valuable information that could be useful to others. Who knows? You may help a fellow student get a summer job this year, which will result in that student opening doors for you next year.

OTHER SOURCES OF EMPLOYMENT LEADS. There are many other sources of information about engineering employers. The classified ads in the newspaper can give you a clue as to who is hiring, even though the positions advertised might not be for you. Your university reference librarian can assist you in finding publications that list employers. The *National Business Employment Weekly* published by the *Wall Street Journal* is good source that should be available in your library. Similarly, there are other publications [8,9] containing good leads.

The North American Industry Classification System (NAICS) discussed in Chapter 2 can be used to research industries that employ engineers. First identify a specific industry (one of the 1,170 industries identified by a five-digit or six-digit NAICS classification code) and choose a product or service that you'd like to be involved in. Next, enter the name of the product or service into an Internet search engine (e.g., http://www.google.com) to identify specific companies that make the product or deliver the service. Go to each of these companies' web page to learn as much as you can about them. Pick one or more of the companies and contact them about employment possibilities. Concentrating your job search on companies you find interesting and know a lot about will give you a distinct advantage.

USING THE INTERNET. The Internet provides an inexhaustible resource for tracking down job opportunities. You can seek help and direction about the best web sites from your career center, or you can do it on your own. Following are just a sample of the many Internet web sites that could help advance your job search. (And you'll be pleased to know that your use of these services is free!)

Monster
http://www.monster.com
Monster.com is one of largest job search web sites on the Internet, claiming over a million job postings at any time and over 41 million resumes in its database. It also features employer profiles, job search and career advice, and links to other career sites.

MonsterTRAK
http://www.monstertrak.monster.com
MonsterTRAK (formerly JobTRAK.com) is the most visited college-targeted job search web site. To access this site, you will need to get a password from your career center. You'll find lots of general job search information on the MonsterTRAK web site, and you will be able to upload your current resume or create a new one on-line. Jobs are posted by type (full-time, summer, part-time), by location, and by discipline.

CareerBuilder
http://www.careerbuilder.com
This site contains engineering job listings, a monthly newsletter, and resume and cover letter advice.

InterEC.NET
http://www.interec.net
This web site is specifically set up for engineers looking for a job over the Internet. The site includes job search resources for specific engineering disciplines, job listings, resume databases, and employment data.

OTHER WAYS TO USE THE INTERNET. You will also find job listings on the web pages of most of the professional engineering societies. For example, go to http://www.asme.org and click on "Career." Then click on "Find a Job."

All the major Internet search engines (e.g., Google, MSN, AltaVista, Yahoo, etc.) can be useful in exploring job opportunities. For example, go to http://www.yahoo.com and click on "Hot Jobs." Or go to http://www.google.com, and enter "job search."

You probably realize by now that a few hours on the Internet can provide you with more employment leads than you will ever be able to pursue. Your challenge will be to select those few that best match your needs. In the next section we tell you how to follow up on those leads.

EXERCISE

Go to http://www.careerbuilder.com. Click on "Find Jobs." Enter keyword "engineering" and conduct a search for a listing of engineering jobs. How many are listed? Redo your search for a specific engineering discipline (e.g., civil engineering) and a specific geographical area (e.g., your state). How many jobs are listed? Scroll through the listing and identify several positions that look interesting to you. Review the detailed job description and qualifications for those positions.

APPLYING FOR POSITIONS

The most straightforward way to follow up on a lead is to call the company and ask for the name and title of the individual in charge of the company's student-hiring program. Send this person a cover letter and resume. In the cover letter, state that you will follow up with a telephone call within two weeks.

Your primary goal should be to get to an interview. An interview will give you the best opportunity to "sell" yourself using the interviewing skills you have developed. Getting an interview is not easy, however. Industry representatives, whether they be in the human relations department or in the engineering line organization, are generally very busy. They have too many candidates for employment, and too little time to evaluate applications.

If you do get an interview, follow the guidelines presented previously in the section on *Developing Your Interviewing Skills*. But you don't have to wait to be invited for an interview. You can take the initiative by arranging an "informational interview."

INFORMATIONAL INTERVIEWS. The informational interview is not a job interview. It is an information-gathering session. In a job interview, the employer is interviewing you. In an informational interview, you are interviewing the employer.

How do you arrange an informational interview? A good way is through networking. Perhaps through a friend or a member of your family you can get the name of an engineering manager at a local company. You then telephone that person, using the name of your friend or family member as a reference, and request 20 to 30 minutes of the person's time to learn about the company and the kind of work done there.

Although personal referrals are helpful, you can arrange informational interviews without them. Any alumnus of your engineering program would very likely be willing to meet with you. Or you can just use the fact that you are an engineering student and would like to learn more about career opportunities in engineering. Your position as a student can get you through more doors than you think. Consider the following ideas.

Student Power

<u>**Power**</u>—"the ability to influence others"—comes to people from at least three sources: (1) their position; (2) their knowledge; and (3) their person. You probably don't realize how much power you derive from your position as a student. You are in an excellent position to "influence others," and you may not even realize it. The basis for this power is very obvious. Almost anyone you would want to influence spent many years in the very position you now hold—i.e., the position of student. And that person most likely has lots of "warm, fuzzy" feelings about that period of his/her life. Even more important, such people realize they owe much of their success to their education.

So when you call an engineering executive and explain, "I'm a first-year engineering student, and need just 15 minutes of your time to ask you a few questions for an important project I am doing," nine times out of ten that person will agree to meet with you. Try it!

To use this newfound *power*, especially if you lack a personal referral, call a local engineering firm and ask to speak to the chief engineer. If you can't reach him or her, you will probably be referred to someone at a lower level. You can then truthfully say that you were referred by the chief engineer's office and would like to meet with that person to learn more about what the company does.

In preparing for the informational interview, make up a list of questions you plan to ask. The following are some examples:

What do you do in your current position?
What are the most satisfying aspects of your work?
What is your educational background?
Which of the courses you took in college have been most useful to you in your career?
What was your first job after graduating from college?
How did you go about getting that position?

How is your company's business picture?
What is the future hiring situation?
How important is it for engineering students to gain engineering-related work experience?
Can you advise me as to how I might get a position that will give me that experience?

Once again, remember that people enjoy helping others and giving advice. They also like to talk about themselves. Recall the story of the coal salesman, Mr. Knaphle, in Chapter 4. By showing that you are interested in other people and want to learn from them, they will become interested in you. You may find that they offer to help you get a summer job without your even asking. Even if they don't, you can always send them an application for employment at a future date.

FOLLOWING UP ON INTERVIEWS

Whether you have a job interview or an informational interview, it is important that you follow up. Always send a thank-you letter. Few people do, so if you do, you will be remembered positively. In your letter, thank the interviewer for his or her time and interest. Be sure to mention some specific information you learned that you found particularly useful. If you are following up on a job interview, express genuine interest in the job opportunity. Conclude by leaving the door open for you to contact them in the future.

7.4 PUTTING SOMETHING BACK

I'm sure you have heard someone say, "I'm not going to vote. My vote doesn't really count." In one way, that view makes sense. After all, with millions of votes cast, one vote isn't really likely to make a difference. But what if everyone held the view that one vote isn't important? Since we can't afford to have everyone decide not to vote, it's not right for one person to do so.

How do you view your relationship with your university or college? Do you feel that you have something to offer your institution? Or do you feel that your contributions are not important—that what one student does cannot really make a difference? I hope you see the parallel with the importance of voting and realize that it is important to "cast your vote" with your university—that you put something back into the institution that is giving you so much.

President John F. Kennedy motivated an entire generation of young Americans when he said:

> ***Ask not what your country can do for you; ask what you can do for your country.***

Ask yourself, "What can I do for my institution?" Doing things that will benefit your institution is a real "win-win" situation. The institution wins because what you do will make it a better place for its students, faculty, and staff. You win in two ways. You will reap direct rewards from experiences gained from what you do. And you will benefit because the quality of your institution will be improved.

Giving to your institution can—and should—continue throughout your lifetime. After you graduate, you will be an alumnus of the university. As an alumnus, you will have the ongoing opportunity to enhance your institution through contributions of both your time and money. To some extent, the value of your education is related to the image others have of your university or college. If the image of your university improves, even many years after you have graduated, the value of your education will be enhanced. So whether you want it or not, you and your university will be permanently linked.

Following are some of the ways you can put something back into your university or college, even as an undergraduate student. Doing many of the things we have already discussed, such as performing well in your classes and becoming actively involved in extracurricular activities, will by their very nature benefit your university. But there are other ways you can contribute:

(1) Providing feedback

(2) Serving as an ambassador

(3) Helping other students

This is not a complete listing, and I'm sure you can think of other ways to give something back to your university. But let's consider these three suggestions for now.

Providing Feedback

You are your institution's primary customer. You know best whether you are getting what you need. You therefore should make every effort to let those in decision-making positions know how the institution is serving its customers. Don't restrict your feedback to negative remarks. Positive feedback can have as much, or even more, value in bringing about positive change than negative feedback.

You will have some formal opportunities to provide this kind of feedback. The best example is when you are invited to complete student opinion surveys about your professors' classroom performance. Please take these surveys seriously. They not only give feedback to your professors that they can use for self-improvement; the results of the student opinion surveys are used in important decisions about tenure, promotion, and merit salary increases. Generally, these surveys consist of a series of numerical questions followed by a place for you to write narrative comments. I strongly encourage you to write detailed comments. As a professor, I found the comments far more informative and useful than the numerical results.

You will undoubtedly have other opportunities to provide feedback about your education and your institution. You may be invited to write letters of support for professors; you may receive surveys designed to measure the overall campus climate; or you may see an invitation to students to meet with the dean or department chair to give feedback. I hope you take full advantage of these and other opportunities to give feedback.

You can also give unsolicited input. Be liberal with positive feedback. As we discussed in Chapter 4, let your professors know when you like the subject or value their teaching. Tell the dean or department chairs about anything you like. People are less receptive to negative feedback, so you should be more selective with negative criticism. But if you really feel that something important is not right, don't hesitate to make an appointment to see the dean or the department chair to air your grievances. If you do, make every effort to present yourself in a tactful, respectful, and rational manner.

Serving as an Ambassador

You are also your university's best ambassador. There are both formal and informal opportunities for you to serve in ambassador roles. Your university may have a formal *ambassadors' organization* of students who represent the university at a variety of events. Such ambassadors conduct special tours; host receptions, dinners, or special events; serve as ushers; or escort distinguished visitors and alumni.

Your university may also have a community service organization similar to the Educational Participation in Communities (EPIC) program. Through this type of organization, you can volunteer for community service assignments in schools, hospitals, community centers, and other human service agencies.

You can create your own ambassador activities as well. Return to your high school or other high schools and speak to teachers and students there on behalf of the university. "Word of mouth" is one of the university's best image-builders. When you speak to anyone off campus, take the view that you are representing the university. Put forth the most positive perspective you are capable of. Keep in mind that any time you "bad mouth" your university, you are diminishing the value of your education. Keep your complaints on campus and tell them to someone who can do something about them.

HELPING OTHER STUDENTS

Can you recall times when others students helped you? What did they do for you? Perhaps they pointed you toward a great teacher, provided you with information about some regulation or campus resource that really benefited you, or gave you some free tutoring that clarified a point you were stuck on.

Don't always be the one who is seeking help from others. Look for opportunities to help other students. Although what you have to offer will increase as you progress through the curriculum, even as a freshman you can help other students. This help can be either informal through contacts you initiate or through work as a volunteer in more structured situations. Volunteer to serve as a computer consultant in the engineering computer lab. Volunteer to work as a peer tutor in your university learning resource center. Or volunteer to work as a peer advisor in special programs for "at risk" students.

You will find that when you help others, you will get more out of it than they do. You will develop your interpersonal communication skills, increase your knowledge, and feel good about yourself for having done it.

SUMMARY

The purpose of this chapter was to introduce you to a number of activities, in addition to your formal academic work, that will broaden and enhance the quality of your education. Through participation in these activities, you will build your interpersonal communication, teamwork, organizational, and leadership skills—skills that will be critically important to your success in your engineering career.

First, we described opportunities for participation in student organizations, particularly those organizations based in the engineering college, and we noted the benefits of such participation. These benefits include establishing relationships with other engineering students, developing your organizational and leadership skills, gaining valuable career information, improving your academic performance, and bolstering your self-esteem by giving something back to your engineering college or community.

Next, we discussed the value of participation in engineering projects such as student design competitions, technical paper contests, Design Clinics, and research projects. These activities require considerable time on your part, but the return can be enormous.

Then we discussed the value of gaining engineering-related work experience through pre-professional employment such as summer jobs, part-time jobs, and cooperative education experiences. We also presented approaches and strategies you can use in conducting successful job searches. Developing job search skills now will be invaluable to you when you seek employment as you near graduation.

Finally, we described several ways you can give something back to your university. As your university's most important "customer," you can provide invaluable feedback. As its ambassador, you can represent it best with external constituencies. And you can be of great help to other students, just as other students have been and will continue to be of help to you.

Taking advantage of the activities described in this chapter takes initiative on your part. Unlike your formal academic work, no one will require you to do them, and no one will check up on whether you do. But the "return on investment" can be even greater than the return you receive from your formal coursework. The activities outlined in this chapter truly offer opportunities for you to take responsibility for the quality of your education.

REFERENCES

1. Remer, Donald S., "Experiential Education for College Students: The Clinic—What It Is, How It Works, and How to Start One," Monograph Series of the New Liberal Arts Program, Research Foundation of the State University of New York, Stony Brook, NY, 1992.
2. Buonopane, Ralph A., "Cooperative Education—Keeping Abreast of New Technologies," *ChAPTER One*, American Institute of Chemical Engineers, May, 1990.
3. Bennett, Scott, *The Elements Of Resume Style: Essential Rules And Eye-opening Advice For Writing Resumes And Cover Letters That Work*, AMACOM Books, June, 2005.
4. Ireland, Susan, *The Complete Idiot's Guide to the Perfect Resume,* Fourth Edition, Alpha Books, February, 2006.
5. Jackson, Tom, "Resumes, Cover Letters, and Interviews," *National Business Employment Weekly*, October, 11-17, 1991.
6. Deluca, Matthew, *Best Answers to the 201 Most Frequently Asked Interview Questions*, McGraw-Hill, New York, NY, 1996.
7. Kador, John, *201 Best Questions to Ask on Your Interview*, McGraw Hill, February, 2002.
8. *Peterson's Hidden Job Market 2000: 2,000 High-Growth Companies That Are Hiring at Four Times the National Average*, Peterson's Guides, Princeton, NJ, 1999.
9. Wallace, Richard (Editor), *Adams Jobs Almanac*, Ninth Edition, Adams Media Corporation, Holbrook, MA, October, 2006.

PROBLEMS

1. Make a list of all the engineering student organizations at your institution. Are you an active member of one or more of these organizations? If not, join the one you are most interested in.
2. Visit the web site of the national engineering society you are most interested in. Locate information on scholarships and awards given to students by the society. Share this information with fellow students in your Introduction to Engineering class. Determine whether you are eligible for one of the scholarships or awards, and if so, apply for it.

3. Determine whether there is a local section of the national engineering society you are most interested in. Find out when the local section holds its meetings, and attend one of them. While at the meeting, try to meet as many of the members as you can. Ask one of them if you can visit them to conduct an "informational interview."
4. Find out if the student chapter you joined is organized to accomplish the five purposes outlined in Section 7.1. If not, suggest that a committee structure be put in place to address any missing purpose (e.g., Social Committee, Professional Development Committee, Personal Development Committee, Academic Development Committee, Service Committee). Volunteer to chair one of these committees and develop a plan for the next year's activities.
5. Visit the office of your university-wide student government. Arrange to meet the student body president and ask him or her whether there are any open committee assignments you could volunteer for.
6. Find out whether your engineering college has participated in any of the engineering student design competitions listed in Section 7.2. If there is one or more, which one(s)? Consider getting involved in the competition.
7. If the answer to Problem 6 is "none," pick the one you are most interested in. Contact the sponsor of the competition to obtain detailed information on the event. Try to persuade your engineering college to participate.
8. Find out if your engineering college has a "Design Clinic" program in which undergraduate students work in teams on real world engineering problems. Find out how you can participate.
9. Find out how many full-time faculty members there are in your college of engineering. Determine which faculty members have funded research projects. Find out how many of them employ undergraduate students to work on their research projects. Make a commitment to seek such an opportunity.
10. Determine whether your engineering college has a formal cooperative education program. (Note: The co-op program may be operated university-wide rather than by each academic unit.) If there is a formal co-op program, visit the co-op office and find out how one applies for a co-op position.

11. Visit your career center. Ask for a list of all companies that interview on campus for engineering graduates. Pick one of the companies and research it through its web site. Write a 500-word essay about the company.

12. Do a personal assessment based on the three factors listed in the section on "How Do You Measure Up?" to find out how well you will do in competing for pre-professional employment positions. If you are not satisfied with your competitiveness, make a plan for improvement. Implement the plan.

13. Acquire one of the books on conducting job searches listed in the references for this chapter (Refs. 3, 4, 6, or 7). Read the book and write a two-page explanation of what you learned. (Note: All of the books are available through http://www.amazon.com).

14. Based on the instructions given in the section on "Preparing a Resume," write your own resume. Ask several people to critique it. These could be fellow students, professors, career center staff, or practicing engineers. Revise your resume based on the input you get, and develop plans to make your resume more impressive (e.g., join and participate in student organizations; find out how to earn academic awards, scholarships, or recognition; map out strategies for landing pre-professional engineering jobs; etc.). Commit to having a resume that is "ready to go" throughout your college years.

15. Write a cover letter seeking a summer job as an engineering aide with the company you selected in Problem 11. Have the cover letter critiqued by several people and revise it until you are satisfied with it. Send the letter and your resume to the company early in the spring term. Follow up on your application as explained in Section 7.3.

16. Get a friend or fellow student to ask you the questions presented in the section on "Developing Your Interviewing Skills." Have the person critique your answers.

17. Prepare a written response to each of the questions presented in the section on "Developing Your Interviewing Skills." Practice your answers and then repeat Problem 16. Did you note any difference?

18. Make up ten additional questions that you think you might be asked in an interview for a summer job. Prepare responses to those questions.

19. Use the Internet to do the following:

    a. Go to http://www.census.gov and click on "NAICS."

    b. Search NAICS to identify an industry (as identified by its five-digit or six-digit NAICS classification code) that you are interested in.

    c. Pick a product or a service in that industry.

    d. Using one of the Internet search engines such as http://www.google.com, enter the name of the product or service and identify the names of companies that make the product or deliver the service.

    e. Pick one of those companies, go to their web page, and learn as much as you can about that company.

    f. Contact the company (either by e-mail or by telephone). Find out if it has summer job positions for engineering students and, if so, how one can apply for such positions.

20. Make a list of ten companies you would like to work for in the summer using the methods outlined in the section on "Identifying Employment Opportunities." Plan a campaign to apply for a pre-professional employment position with each of them.

21. Pick three of the companies you found in Problem 19. Either through networking or through a telephone call to the chief engineer of each company, identify a person with whom you can conduct an *Informational Interview*. Arrange these interviews. Write a critique discussing how each interview went. Don't forget to send a "thank you" letter after each interview.

22. Evaluate the list of questions presented in the section on *Informational Interviews*. Rate each question on a scale of 0 to ten. Think up five additional questions that you might ask. Rate those questions. Take the total list of 15 questions and select ten you would feel comfortable asking a practicing engineer. Order the questions in what you feel is most logical. Use this list when you conduct any *Informational Interviews*.

23. Make up a list of questions you would ask a professor during an *Informational Interview*. Pick one of the engineering professors and arrange a 20 to 30 minute meeting with him or her. Write a critique of the interview.

24. Go to the MonsterTRAK web site http://www.monstertrak.com and determine whether your university is registered. If it is, contact your career center and ask for the "password" needed to access MonsterTrak job listings. Return to the MonsterTrak web site and locate any part-time or summer job listings in engineering in your state.

25. Access the web site: http://www.interec.net. Explore the jobs listed for several engineering disciplines of interest to you. Write a two-page summary about the jobs you found there.

26. Go to the "CareerBuilder" web site at http://www.careerbuilder.com. Explore the job search resources available there. Prepare a three-minute oral presentation for your Introduction to Engineering classmates describing what you learned.

27. Write down ten positive features of your university. Rank them in order of importance. Pick ten different people (students, faculty members, department chair, dean). Tell each person about a feature on your list. How did they respond?

28. Find out whether your engineering college or university has any service-oriented clubs. Write a one-page description of what one of the service organizations does. Plan to participate in the service club for at least one term during your undergraduate years.

CHAPTER 8
Orientation to the Engineering Education System

Destiny is not a matter of chance,
it is a matter of choice.

William Jennings Bryan

INTRODUCTION

The purpose of this chapter is to orient you to the engineering education system of which you are a part. If you are to take full advantage of your education, it is important that you understand how that educational system works. And there is an added bonus. As you better understand the engineering education system and make that system work for you, you will develop the ability to understand other systems you will encounter in the future, and you will gain the skills needed to make those systems work for you as well.

First, we will provide an overview of how engineering education is organized in the United States and how engineering programs are organized within colleges and universities.

Next, we will consider the important role community colleges play in engineering education. You may currently be a community college student or you may have transferred from a community college to a four-year institution. Completing your first two years of engineering study at a community college can have distinct advantages. And even if the community college does not have a formal engineering program, you should be able to complete the majority of the lower-division engineering requirements there.

Then we will provide you with an overview of the engineering education system. To do this, we will use the criteria the Accreditation Board for Engineering and Technology (ABET) requires all engineering programs to meet. By understanding these criteria, you will gain insight into the key elements that comprise your engineering program: students, educational objectives, program outcomes, curriculum, faculty, facilities, and institutional support and financial resources.

Next, we will examine important academic regulations, policies, and procedures in three areas: (1) academic performance; (2) enrollment; and (3) student rights. Knowing these regulations, policies, and procedures at your institution will enable you to make optimal use of the educational system.

Finally, we will consider opportunities for education beyond the B.S. degree in engineering—both graduate study in engineering and post-graduate study in other closely related disciplines. The benefits of pursuing an M.S. or Ph.D. degree in engineering will be discussed. Among those benefits are the advanced technical knowledge you will derive from the additional coursework and the research skills you will gain by completing a thesis or dissertation under the close supervision of a faculty advisor. Last, we will describe opportunities to pursue post-graduate study in business administration, law, and medicine.

8.1 Organization of Engineering Education

According to statistics complied by the U.S. Department of Education [1], in 2004 more than 17 million students were enrolled in 4,216 colleges and universities in the United States. Sixty percent were full-time students and 40 percent studied part-time. Of the 17.1 million students, 86 percent were enrolled in undergraduate study and 14 percent were engaged in graduate and professional study. Among the 4,216 colleges and universities are 1,683 two-year institutions, 639 public four-year institutions, and 1,894 private four-year institutions.

Overview of Engineering Education in the U.S.

Of the 2,533 four-year colleges and universities (639 public and 1,894 private) in the U.S., only 352 (13.9 percent) have accredited engineering programs [2]. As we learned in Section 2.4, these 352 institutions offer a total of 1,495 accredited engineering programs—an average of slightly over four programs per institution. The number is not uniform, however. A few institutions offer as many as 14 different engineering programs, while others offer only one program.

Each of the nation's 1,495 accredited engineering programs is evaluated by the Accreditation Board for Engineering and Technology (ABET). Gaining ABET accreditation is extremely important. It is unlikely that any program could survive without being accredited.

To earn that accreditation, a program must meet high standards of quality for students, faculty, curriculum, administration, facilities, and institutional support and resources. Each program must also demonstrate that its graduates have acquired specific knowledge and skills; and each program must have a "continuous improvement" process in place to further develop and improve its quality. We will discuss the accreditation process in more detail in a later section.

ORGANIZATION OF THE ENGINEERING UNIT

Each engineering program is administered by an engineering department (e.g., Department of Civil Engineering). Generally, a department administers only one engineering program, but it is not uncommon to find two or three programs administered by a single department (e.g., Department of Mechanical and Aerospace Engineering). At the head of each department is the *Department Chair* or *Department Head.*

The engineering departments at a university are generally organized into a "school" or "college" of engineering, headed by the *Dean.*

Non-engineering departments may also be part of the school or college in which the engineering departments reside. Computer science, engineering technology, and industrial technology are the three most common of these. For example, the engineering departments and the computer science department could be organized into a College of Engineering and Computer Science. At some small institutions, the engineering programs may be combined administratively with the mathematics and science departments to form a College of Science and Engineering.

POSITION OF THE ENGINEERING UNIT IN THE UNIVERSITY

The engineering college is only one of several schools or colleges on a university campus. Other colleges might include the College of Business, the College of Arts and Letters, the College of Natural Science, the

College of Education, and the College of Health and Human Services. All of the colleges on a campus are organized into the "Academic Affairs" unit headed by the *Vice President* or *Vice Chancellor for Academic Affairs*. (Often this person also carries the title of *Provost*.)

The Vice President or Vice Chancellor for Academic Affairs reports to the *President* or *Chancellor* of the university. The president oversees the entire university. In addition to academic affairs, the President or Chancellor is also responsible for such ancillary operations as fiscal management, facilities management, information resources management, student affairs, institutional advancement, and auxiliary services.

The organization of the academic side of the university, from the engineering department chair to the president, is shown below.

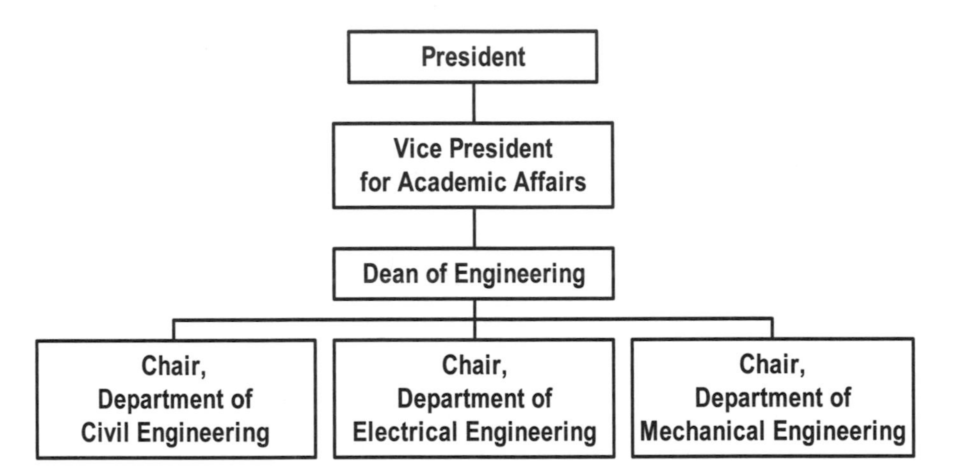

8.2 Community College Role in Engineering Education

Community colleges represent a major part of the nation's higher education system. As previously indicated, there are 1,683 community colleges in the U.S. Forty-five percent of the nation's 14.7 million undergraduate college students are enrolled in community colleges [1].

Community colleges are a very important part of the overall engineering education system. Many engineering graduates started their engineering study at a community college, and then transferred to a four-year institution to complete their B.S. degree in engineering. In fact, a recent study by the National Science Foundation [3] indicated that 40 percent of engineering graduates in 1999 and 2000 had attended a community college at some time.

Many community colleges offer lower-division engineering or pre-engineering programs that allow students to complete all of their lower-division requirements and then transfer to a four-year institution to complete their upper-division engineering requirements. Some community colleges offer Associate of Science (A.S.) degrees in engineering, but only a small fraction of engineering students complete the requirements for that degree. For example, in 2003-04, only 2,737 A.S degrees in engineering were awarded in the U.S. compared to 63,558 B.S. degrees in engineering [1].

Where community colleges do not have formal engineering programs, students can still complete about 70 to 80 percent of the lower-division engineering requirements by taking the required calculus, chemistry, physics, and lower-division general education courses—courses offered by all community colleges.

ENGINEERING TECHNOLOGY

Community colleges also offer engineering technology programs. *Engineering technology* is a field that is closely related to engineering, but has a more practical focus. The difference is explained by Lawrence J. Wolf, former president of Oregon Institute of Technology [4]:

> "Engineering technology draws upon the same body of knowledge as engineering, but centers more heavily on the applications related to manufacturing, testing, construction, maintenance, field service, and marketing."

Although the opportunity exists for engineering technology students to transfer to four-year institutions to pursue their B.S. degree in engineering technology, the majority of engineering technology students terminate their education with the A.S. degree. This is evidenced by the fact that 36,915 A.S. degrees in engineering technology were awarded in 2003-04 by both two-year and four-year institutions; whereas, only 14,669 B.S. degrees in engineering technology were awarded in the same year by four-year institutions [1].

Some A.S. and most B.S. degrees in engineering technology are accredited by the Accreditation Board for Engineering and Technology (ABET). A listing of these programs by discipline, degree level, and state or region can be found on the ABET web site at:

http://www.abet.org/accredittac.asp

An excellent source of information about engineering technology is Chapter 2, "The Field of Engineering Technology" from Stephen R.

Cheshier's text *Studying Engineering Technology: A Blueprint for Success* [5]. The chapter can be accessed on the web at:

http://www.discovery-press.com/catalog/studyent/Chap2.htm

ARTICULATION AND COURSE SELECTION

Community colleges having formal engineering programs generally develop *articulation agreements* with four-year institutions in their geographic area. The articulation agreements guarantee students that specific courses taken at the community college will be transferable to the four-year institution. Articulation agreements can be on a course-by-course basis, or they can apply to the full lower-division program ("2+2" articulation agreements).

Where such articulation agreements exist, they can provide you with the "road map" you need to plan your lower-division coursework. Particular attention should be provided to the selection of the general education courses (humanities, social sciences, communication skills) you take at your community college. The general education requirements for an engineering degree at a four-year institution may differ from those for other majors. Unless you are careful, you might find later that you have taken courses that will not be counted among the requirements for your B.S. degree in engineering.

One additional point regarding general education courses. You may receive advice or encouragement to complete most or all of your general education requirements at your community college. I would suggest that you ignore this advice and save about half of your general education courses to balance your technical course load during your last two years of engineering study at the university. By doing so you'll avoid the daunting task of having to take full loads of only advanced-level engineering courses during your junior and senior years of engineering study.

ADVANTAGES OF STARTING AT A COMMUNITY COLLEGE

High school graduates have the choice of starting their engineering study at a community college or at a four-year institution. There are advantages and disadvantages associated with either choice. The choice usually depends on a student's high school record, financial situation, and personal needs. The following is a discussion of the advantages of starting at a community college.

If your record from high school does not qualify you for admission to the university of your choice, by attending a community college you, in effect, get a second chance. By building a strong academic record at the community college, you will then be able to transfer to the four-year institution of your choice. If you need to bring your skills in mathematics, science, and English up to the university level, generally a greater range of developmental courses in these areas are available at a community college than at a four-year institution.

Your time at a community college could very well provide you with the luxury of taking some extra courses designed to give you one or more "saleable skills." Examples of these skills areas are computer-aided-drafting, surveying, web design, computer programming, machine tool operation, and electronic trouble shooting. The value of taking courses in such areas—courses that may not be required as part of your engineering curriculum—is that they provide you with skills that can help you find engineering-related jobs during your tenure as a student; skills that may also come in handy later during your professional engineering career.

Lower cost is another advantage of attending a community college. A student who lives at home can meet his or her community college educational expenses by working as little as ten to 15 hours per week.

Finally, the community college environment lies somewhere between the warm, friendly, small school environment you experienced in high school and the less friendly, large school environment you will find at many major universities. Hence, you may find that a community college is a place that will provide you a more supportive learning environment in which to mature, grow, and develop before transferring to a four-year institution.

Applicability of This Book to Community College Students

If you are a community college engineering major, you will find that the concepts put forth in this book will apply directly to your situation. The first two years of engineering study at a community college are similar in virtually all regards to the first two years of engineering study at a four-year institution.

There is one exception. Various co-curricular activities discussed in Chapter 7—such as research projects, engineering student organizations, and engineering student design projects—are generally more available to students at four-year institutions than at community colleges. However, involvement in such activities is more likely to occur during a student's junior and senior years, so you won't miss too much.

In any case, if you are a community college student, the sections in this book that apply primarily to four-year institutions should give you a useful preview of what you can expect when you transfer to one.

8.3 THE ENGINEERING EDUCATION SYSTEM

The ABET accreditation process provides a useful framework for understanding the engineering education system. Once every six years, a team comprised of practicing engineers and engineering educators representing the **Accreditation Board for Engineering and Technology** conducts a three-day visit to your institution to evaluate all aspects of your engineering program. The purpose of the evaluation is to ensure that the engineering program meets or exceeds specific criteria in eight areas.

Criterion 1	**Students**
Criterion 2	**Program Educational Objectives**
Criterion 3	**Program Outcomes and Assessment**
Criterion 4	**Professional Component**
Criterion 5	**Faculty**
Criterion 6	**Facilities**
Criterion 7	**Institutional Support and Financial Resources**
Criterion 8	**Program Criteria**

Programs that meet or exceed all criteria are accredited for a six-year period. Programs with minor deficiencies will either be revisited in two years or required to write a report documenting progress in correcting the deficiencies. Serious deficiencies can result in the program being put on probation, which could lead to the loss of ABET's accreditation.

The following sections present each of the eight criteria specified in ABET's "Criteria for Accrediting Engineering Programs." [6]. You can find additional information about the accreditation process and criteria on the ABET web site: http://www.abet.org.

CRITERION 1 - STUDENTS

The quality and performance of the students and graduates are important considerations in the evaluation of an engineering program. The institution must evaluate student performance, advise students regarding curricular and career matters, and monitor student's progress to foster their success in achieving program outcomes, thereby enabling them as graduates to attain program objectives.

The institution must have and enforce policies for the acceptance of transfer students and for the validation of courses taken for credit elsewhere. The institution must also have and enforce procedures to assure that all students meet all program requirements.

CRITERION 2 - PROGRAM EDUCATIONAL OBJECTIVES

Program educational objectives are broad statements that describe the career and professional accomplishments that the program is preparing graduates to achieve.

Each engineering program for which an institution seeks accreditation or re-accreditation must have in place:

a. detailed published educational objectives that are consistent with the mission of the institution and these criteria
b. a process based on the needs of the program's various constituencies in which the objectives are determined and periodically evaluated
c. an educational program, including a curriculum that prepares students to attain program outcomes and that fosters accomplishments of graduates that are consistent with these objectives
d. a process of ongoing evaluation of the extent to which these objectives are attained, the result of which shall be used to develop and improve the program outcomes so that graduates are better prepared to attain the objectives.

CRITERION 3 - PROGRAM OUTCOMES AND ASSESSMENT

Program outcomes are statements that describe what students are expected to know and be able to do by the time of graduation. These relate to the skills, knowledge, and behaviors that students acquire in their matriculation through the program.

Each program must formulate program outcomes that foster attainment of the program objectives articulated in satisfaction of Criterion 2 of these criteria. There must be processes to produce these outcomes and an assessment process, with documented results, that demonstrates that these program outcomes are being measured and indicates the degree to which the outcomes are achieved. There must be evidence that the results of this assessment process are applied to the further development of the program.

Engineering programs must demonstrate that their students attain:

a. an ability to apply knowledge of mathematics, science, and engineering
b. an ability to design and conduct experiments, as well as to analyze and interpret data
c. an ability to design a system, component, or process to meet desired needs within realistic constraints such as economic, environmental, social, political, ethical, health and safety, manufacturability, and sustainability
d. an ability to function on multi-disciplinary teams
e. an ability to identify, formulate, and solve engineering problems
f. an understanding of professional and ethical responsibility
g. an ability to communicate effectively
h. the broad education necessary to understand the impact of engineering solutions in a global, economic, environmental, and societal context
i. a recognition of the need for, and an ability to engage in life-long learning
j. a knowledge of contemporary issues
k. an ability to use the techniques, skills, and modern engineering tools necessary for engineering practice.

In addition, an engineering program must demonstrate that its students attain any additional outcomes articulated by the program to foster achievement of its education objectives.

CRITERION 4 - PROFESSIONAL COMPONENT

The *professional component requirements* specify subject areas appropriate to engineering but do not prescribe specific courses. The faculty must ensure that the program curriculum devotes adequate attention and time to each component, consistent with the outcomes and objectives of the program and institution. The professional component must include: (Note: "One year" is 32 semester units or 48 quarter units)

a. one year of a combination of college level mathematics and basic sciences (some with experimental experience) appropriate to the discipline
b. one and one-half years of engineering topics, consisting of engineering sciences and engineering design appropriate to the student's field of study. The engineering sciences have their roots

in mathematics and basic sciences but carry knowledge further toward creative application. These studies provide a bridge between mathematics and basic sciences on the one hand and engineering practice on the other. Engineering design is the process of devising a system, component, or process to meet desired needs. It is a decision-making process (often iterative), in which the basic sciences, mathematics, and the engineering sciences are applied to convert resources optimally to meet these stated needs.

c. a general education component that complements the technical content of the curriculum and is consistent with the program and institution objectives.

Students must be prepared for engineering practice through the curriculum culminating in a major design experience based on the knowledge and skills acquired in earlier course work and incorporating appropriate engineering standards and multiple realistic constraints.

CRITERION 5 - FACULTY

The faculty is the heart of any educational program. The faculty must be of sufficient number; and must have the competencies to cover all of the curricular areas of the program. There must be sufficient faculty to accommodate adequate levels of student-faculty interaction, student advising and counseling, university service activities, professional development, and interactions with industrial and professional practitioners, as well as employers of students.

The program faculty must have appropriate qualifications and must have and demonstrate sufficient authority to ensure the proper guidance of the program and to develop and implement processes for the evaluation, assessment, and continuing improvement of the program, its educational objectives and outcomes. The overall competence of the faculty may be judged by such factors as education, diversity of backgrounds, engineering experience, teaching experience, ability to communicate, enthusiasm for developing more effective programs, level of scholarship, participation in professional societies, and licensure as Professional Engineers.

CRITERION 6 - FACILITIES

Classrooms, laboratories, and associated equipment must be adequate to accomplish the program objectives and provide an atmosphere conducive to learning. Appropriate facilities must be available to foster faculty-student interaction and to create a climate that encourages professional development and professional activities. Programs must provide opportunities for students to learn the use of modern engineering tools. Computing and information infrastructures must be in place to support the scholarly activities of the students and faculty and the educational objectives of the program and institution.

Criterion 7 - Institutional Support and Financial Resources

Institutional support, financial resources, and constructive leadership must be adequate to assure the quality and continuity of the engineering program. Resources must be sufficient to attract, retain, and provide for the continued professional development of a well-qualified faculty. Resources also must be sufficient to acquire, maintain, and operate facilities and equipment appropriate for the engineering program. In addition, support personnel and institutional services must be adequate to meet program needs.

Criterion 8 - Program Criteria

Each program must satisfy applicable Program Criteria. Program Criteria provide the specificity needed for interpretation of the basic level criteria as applicable to a given discipline. Requirements stipulated in the Program Criteria are limited to the areas of curricular topics and faculty qualifications. If a program, by virtue of its title, becomes subject to two or more sets of Program Criteria, then that program must satisfy each set of Program Criteria; however, overlapping requirements need to be satisfied only once.

Reflection

Reread the eight ABET criteria presented in the previous section. Make up a list of questions about issues you don't completely understand or about things you would like to know more about. As examples. What are the "program educational objectives" for our engineering program? Do we have "program outcomes" beyond the ABET (a) – (k) outcomes? What is meant by "design a system for sustainability"? What "modern engineering tools" will I learn to use? What course or courses addresses the ABET requirement for a "major design experience"? As you have the opportunity, ask your professors, your department chair, or your dean these questions.

Conclusion

I'm sure you'll agree that these criteria are comprehensive and are likely to result in the "continuous improvement" of your engineering program. By understanding them, you will better understand the engineering education system. As you progress through your engineering studies, you may want to do your own evaluation as to how your engineering program measures up to these standards.

8.4 ACADEMIC ADVISING

ABET *Criterion 1* states that "The institution must . . . advise . . . students with regard to curricular and career matters," and *Criterion 5* states that "There must be sufficient faculty to accommodate adequate levels of . . . student advising and counseling."

Academic advising, including both curricular and career advising, is extremely important. I hope you are studying in an engineering college in which the engineering faculty take academic advising seriously. Many students are.

On the other hand, you may not be getting the quality and quantity of academic advising you need. Unfortunately, engineering faculty sometimes neglect their advising responsibilities in favor of the demands of teaching and research. According to Phillip C. Wankat, professor of engineering at Purdue University [7]:

> "Probably the most neglected arca in engineering education is advising, and certainly this is the area where students show the least satisfaction."

Wankat's statement has been borne out by my personal experience.

A Personal Anecdote

Often when I visit universities, I ask the dean of engineering, "How is your advising system?" The dean usually tells me something like this: "We have a great advising system. Each student is assigned a faculty advisor and meets with that advisor each term to plan the student's course program for the next term. In the advising session, the advisor reviews the student's past performance, works out the student's course program for the next term, and gives the student any needed career guidance."

Then I ask students, "How does the advising system work here?" More often than not students will tell me, "I leave my advising form with the department secretary, who gets my advisor to sign it, and I pick it up the next day."

QUALITY OF ADVISING CAN BE A PROBLEM

The absence of academic advising is not the only problem. An equally serious problem is bad advising.

One area of bad advice to look out for comes from faculty who believe that you "haven't measured up" unless you graduate in four years. Such advisors will insist that you take 16-18 units, whether this is best for you or not. These faculty can fail to account for the fact that you may be working 20 hours a week, or might have been out of school for a few years and need to start slowly to work up to full speed.

Another Personal Anecdote

> I recall the academic advising I received when I first began my Ph.D. program. I had been working in industry for five years and had to readjust to the demands of academic work. I was assigned an advisor who told me, "Take Dr. Johnson's course. Prove yourself by doing well in that course, and you'll have no problem from then on." Little did I know that "Dr. Johnson's course" was the capstone course in my field that brought together all that would be learned by completing all of the department's graduate courses. Try as hard as I could, I just couldn't handle the course. I dropped it after the midterm exam, and from then on Dr. Johnson had me pegged as a poor student.

Bad advice can also come from advisors who have inaccurate, out-of-date information about the curriculum or lack information about various rules and regulations that affect your academic status. I have a constant stream of students telling me things like the following: "My advisor told me I could try out this course and drop it later." "My advisor told me it would be okay to take 20 units." "My advisor told me that ENGR 322 has been eliminated from the curriculum." Sometimes I can remedy the situation; other times I can't. Don't forget, ***ignorance of the law is no excuse!***

TAKE PERSONAL RESPONSIBILITY FOR GETTING PROPER ADVISING

My recommendation to you is that you take personal responsibility for getting proper academic advising. After all, who suffers when you fail to be advised or get bad advice? You do!

There are several possible sources for academic advising: professors, advising staff, and other students. You can even be your own advisor for matters such as identifying courses you need to take, drawing up a workable schedule for a term, and so on. But you will still need sound academic and career advice from others.

To find a good advisor, first make sure you understand how the advising system is structured at your institution. At some institutions, advising is mandatory; whereas, at others it is optional. One department may assign students individual faculty advisors; another department may have a principal faculty advisor who advises all first-year students. Some engineering schools have advising centers where professional staff do the advising. Regardless of how the advising system is structured at your institution, find out how it works and then take full advantage of it.

If you are assigned an academic advisor, whether a faculty member or professional staff member, you should meet with that person at least once each term when you plan your courses for the next term. An advising session will give you feedback on your academic performance, answer any questions you might have about academic policies or regulations, help you work out your course program for the next term, and provide you with career information.

Fellow students can be good sources of information as well. Students can be helpful in directing you to the best teachers. One warning, however. Just because one student likes a professor, that doesn't mean you will. Professors are not just *good* or *bad*, they are also *hard* or *easy*. Sometimes when a student says that Professor "X" is good, he really means that Professor "X" is easy. I hope you will seek out professors who are good teachers, but also set high standards of performance.

Any advice you get should be tempered with your own judgment and information you can gain from sources such as your institution's catalog, schedule of classes, web page, or student handbook. These sources contain an enormous wealth of information. But you won't get that information unless you use them.

The ideal advisement arrangement is a combination of all sources. As discussed in Chapter 1, you should develop a road map that lays out the courses you plan to take each term throughout your undergraduate years. Share this map regularly with your academic advisor and fellow students. Check it against the four-year curriculum outlined in your school's catalog. Based on all this input—plus your own—follow that road map or revise it whenever it's appropriate until you graduate.

REFLECTION

Reflect on the academic advising you have received thus far. Was it mandatory or optional? Was the advisor a professor? Or a student services staff member? During the advising session, did you discuss your past academic performance? Did you work out your course program for the next term? Did you discuss your long-term career goals? Were you satisfied with the quality of your advising? What could you do to improve the quality of advising you receive in the future?

8.5 ACADEMIC REGULATIONS

It is also important to understand your institution's many academic regulations, policies, and procedures. Not knowing about some of these can hurt you; knowing about others can help you. You can find much of this information in the university catalog or on your university's web page.

I'm not sure exactly why, but I have always been a person who has been able to get the most out of systems. Here's an example.

One Last Personal Story

As I was about to complete my B.S. degree, I realized that I could finish my M.S. degree in seven months (a summer and a semester) by taking advantage of three regulations most students had never heard of:

(1) Senior-level courses beyond those needed to meet the B.S. degree requirements could be applied toward an M.S. degree.

(2) Students in the last semester of undergraduate study could take two graduate-level courses toward their M.S. degree.

(3) M.S. students could petition to enroll in more units than the rules allowed for a student with a full-time graduate assistantship.

I knew that my GPA was marginal for admission to graduate school, so I met with the professor in charge of graduate admissions and persuaded him that my junior and senior year grades justified giving me a chance.

I couldn't even start to tell you all the many ways my career has been enhanced because I stayed for those seven months and completed my M.S. degree.

I hope my story convinces you that you will benefit from understanding your institution's academic regulations, policies, and procedures. By learning them, you might be able to accomplish things that you would not otherwise even think of.

The following sections give brief overviews of important regulations, policies, and procedures that you should know about. These are divided into three categories: (1) academic performance; (2) enrollment policies; and (3) student rights.

ACADEMIC PERFORMANCE

There are a number of regulations, policies, and procedures that affect your overall academic performance. First and foremost are policies related to your grade point average and the way it is calculated. But there are other policies and procedures—such as whether you are allowed to take courses on a *credit/no credit* basis, how "incompletes" are handled, "repeat grade" policies, opportunities for "academic renewal," and credit by examination policies—that if used optimally can help you build a strong grade point average.

GRADE POINT AVERAGE. Your success as a student will be measured in large part by your *grade point average* (*GPA*). I can assure you from personal experience that grades are important. Unlike other factors that are qualitative and difficult to evaluate, your grade point average is quantitative and therefore is likely to get more emphasis than it really deserves.

When I interviewed for the position of dean of engineering at Cal State L.A., I was asked to submit transcripts of all my college work, and I had completed my B.S. degree twenty-three years before! When you interview for your first job, you may or may not be asked to submit transcripts, but you will assuredly be asked about your grade point average. If your GPA is below a certain level, some employers will eliminate you from consideration solely on that basis. Whether this practice is fair doesn't matter; it is a reality you have to face.

Most colleges and universities operate on a 4.0 grade point system as follows:

Grade Symbol	Explanation	Grade Points/Unit
A	Outstanding	4
B	Very Good	3
C	Average	2
D	Barely Passing	1
F	Failure	0

Many universities give plus and minus grades as well. This makes it easier for faculty to grade. Deciding between an *A* and a *B* or a *B* and a *C* in borderline cases can be a difficult decision for faculty. Having *A*- and *B*+ or *B*- and *C*+ as options makes assigning grades a lot easier, while giving a more accurate assessment of your performance.

Your total *grade points* are computed by summing up the product of the credit hours for each course times the grade points per unit corresponding to the letter grade you receive. Your *grade point average* is computed by dividing the total grade points by the total number of units taken.

One last point about your GPA. It's very important that you get off to a good start. Once you have several years behind you, it's very difficult to change your GPA. If you get your GPA off to a bad start, it'll be very difficult to raise it. But the converse is true as well. If you establish a good GPA early on, it's difficult to pull it down.

REFLECTION

One of Stephen Covey's "habits" from his excellent book *Seven Habits of Highly Effective People* is "Begin with the end in mind." As you begin your engineering study, you might want to have in mind that as you near completion of your engineering degree you will be sitting in an interview room, perhaps at your university's career center, being interviewed for a job you would love to have. At some point in that interview, you are certain to be asked "By the way. What's your grade point average?" How are you going to feel if you respond "3.5"? What about "3.0"? What about "2.5" or even lower? Now is the time to do the things that will make that future event a positive experience for you. This is the time to "Begin with the end in mind."

Credit/No Credit. Many universities offer students the opportunity to take courses on a *credit/no credit* (CR/NC) basis. Courses taken CR/NC do not enter into the calculation of your grade point average. Generally, major requirements cannot be taken CR/NC, and the number of units that can be taken on this basis is limited. The benefit of this option, if available, is that it allows you to take courses outside of your areas of strength without the risk of lowering your GPA.

Incompletes. When you are unable to complete a course for justifiable reasons (e.g., illness, family crisis, or job change), you probably can request a grade of *incomplete (I)* from your professor. Generally, the incomplete must be made up within a certain time period. The additional time, however, provides you the opportunity to achieve a higher level of mastery in the course than if you tried to complete it in the midst of a personal crisis.

Repeat Grade Policy. Your university may allow you, under specific conditions, to repeat a course and to count only the higher of the two grades you receive in your grade point average. Generally, you are only allowed to take advantage of this regulation for a limited number of courses. Some universities only allow you to repeat courses in which you have received a grade of *D* or *F*. At other universities, you can even repeat a course to raise a grade of *C* to a *B* or *A*. Check your campus regulations on this.

Academic Renewal. Your university may have a policy that allows students to remove one or more entire terms of coursework from their academic record. Generally, this can only be done under very restrictive circumstances. The policy is designed to "forgive" students who had one or two terms in which their academic performance was extremely low and not representative of what they are capable of. Once again, check your campus policies.

Credit by Examination. Most universities permit students to challenge courses by examination. This is not a "free ride" because whatever grade you receive on the examination, including an *F,* is generally averaged into your grade point average.

Positive and Negative Recognition for Academic Performance

Based on your grade point average, you are eligible for both positive and negative recognition. Negative recognition involves probation and disqualification. Perhaps the most positive recognition you can receive is to be granted your B.S. degree. Other positive recognitions include the Dean's List and Graduation with Honors.

PROBATION. If your grades fall below a certain level, you will be placed on probation. Being placed on probation is a serious warning and indicates that unless your academic performance improves, you will be disqualified. Some universities require that students who go on probation receive mandatory academic advising and/or reduce the number of units they take.

DISQUALIFICATION. Continued poor academic performance will lead to disqualification. "Flunking out" is no fun, and should be avoided at all costs. Policies for reinstatement following disqualification vary from one institution to the next. Some institutions reinstate students immediately following a first disqualification; whereas others require students to drop out of school for a period of time. If you are disqualified a second or third time, you could be permanently barred from the university.

DEAN'S LIST. On the opposite end of the spectrum is the *Dean's List.* This is a very prestigious honor awarded each term to students who achieve a certain level of excellence. Check your university's requirements, but *Dean's List* status generally goes to full-time students whose grades are in the top five percent of students in their major.

GRADUATION REQUIREMENTS. To graduate, you must complete all course requirements for your major with at least a 2.0 (*C* average) overall grade point average. Your university may also require that you have at least a 2.0 grade point average in certain categories of courses, such as all courses in your major field, all general education courses, and all courses attempted at your university. Other typical graduation requirements could include a time limit on courses taken and evidence of skills acquisition such as having to pass an "exit" writing proficiency exam.

GRADUATION WITH HONORS. One of the top recognitions you can receive as a student is to graduate with honors. There are generally three levels of honors: (1) Cum Laude (top 5%); (2) Magna cum Laude (top 3%); and (3) Summa cum Laude (top 1%). Receipt of these honors is usually designated on your diploma and permanent transcript.

ENROLLMENT POLICIES

Every university has a number of regulations, policies, and procedures related to enrollment. These range from how you go about selecting or changing your major to how you register for your classes. Some of the most important of these policies are outlined below.

<u>SELECTING YOUR MAJOR</u>. The procedure by which students select their major differs among institutions. Some require students to designate an engineering discipline during their initial application process. Others will admit students as *engineering (undecided),* and students must apply after their first or second year for admission into a specific discipline based on their academic performance.

Selecting an engineering discipline can be difficult. Your selection should be based on such factors as your aptitude, interest, and employment opportunities—factors about which you may have limited information. At universities where engineering programs have a common lower-division core for all disciplines, the decision can be postponed until the junior or even senior year. For programs that have a highly specialized curriculum requiring an early decision, you may be forced into a decision before you have adequate information.

My advice is to postpone your decision as long as your institution will permit. As you progress through the curriculum, you will be in a better position to choose because of what you learn from your coursework, pre-professional employment experiences, and discussions with students, professors, and practicing engineers.

<u>CHANGING YOUR MAJOR</u>. Don't feel that you must stay with your initial choice of major. As indicated in the previous section, you will gain insights along the way that will enable you to make more judicious decisions about your major. I started out as an electrical engineering major. I didn't like my first "Circuit Theory" course, so I changed to aeronautical engineering. Midway through a course in "Wing Design," I realized that this major was too specialized for me, so I changed to mechanical engineering. Mechanical engineering has turned out to be a great choice for me. My philosophy is summarized by the thought that:

> ***If you get to a <u>good</u> end point, then the path that took you there must have been a <u>good</u> path.***

Don't be worried if you're not sure about what you want to do. View your college years as a chance to explore the possibilities with the purpose of finding out what you like. Take advantage of that opportunity.

One warning. You should check out the procedures for changing your major from one engineering discipline to another. Some disciplines may be easy get into, while others may be oversubscribed and difficult to enter. Don't assume you will be able to change to whatever major you choose.

DOUBLE MAJORS. You can elect to have more than one major. Students with *double majors* must complete all requirements for both majors and receive two bachelor's degrees when they graduate. However, completing the graduation requirements for two majors generally takes at least one full additional year. The extra time can be reduced by choosing a second major that has a great deal of curricular overlap with the first major. For an engineering student, second majors that fit well are those in related areas such as mathematics, physics, or computer science. I wouldn't advocate a double major. In the additional year you would spend to fulfill the requirements for the second major, you could complete a master's degree. An M.S. degree would probably be more beneficial to you than a second bachelor's degree.

MINORS. You can also elect to have a minor field. A minor can offer you most of the benefits of a second major while requiring less additional coursework. A typical minor might require about 12 semester units (18 quarter units) of courses. The minor can be used to strengthen your preparation in an area related to your major (e.g., mathematics, physics, biology) or to gain breadth in a completely unrelated area (e.g., music, philosophy, creative writing).

REGISTRATION. The process of registering for your courses is extremely important. Through the registration process, you can ensure that you get the courses you need, the best instructors, and a workable schedule. Not getting the courses you need can impede your progress, particularly in cases where the courses are prerequisites for future courses. Having the best instructors can also have a major impact on your academic success. And having a good schedule can ensure that you have adequate time for studying and other commitments.

At most institutions, registration is done either on-line or by telephone. You will generally receive a time after which you are allowed to register based on a system established by your registrar's office. For example, the system might have new freshmen register first, followed by seniors, juniors, sophomores, and continuing freshmen. If you are given a low-priority time slot, which will make it difficult to get your classes, you may be able to do something about it. Perhaps the registrar needs volunteers to work at some part of the orientation or registration process and in return allows those students to register early. Often, athletes, band members, and student government leaders are given priority registration. Research all the ways to get priority registration and see if you can qualify for one.

DROP/ADD POLICY. You should become fully versed in your university's *drop/add* policy. You can probably add a course until the end of the second or third week of the term, but if you don't start attending from the beginning, you may have great difficulty catching up. Generally, you can drop a course without any penalty up to a specific date. After that, it becomes more and more difficult to do.

LEAVE OF ABSENCE/WITHDRAWAL. If you decide to leave the university, either for a temporary period or permanently, be sure to follow official procedures. It is generally easy to gain approval for a leave of absence for purposes such as other professional or academic opportunities, travel or study abroad, employment related to your educational goals, field study, medical problems, or financial need. Even if you think you want to leave permanently, don't "burn your bridges." Your situation may change, and you may want to return at some point in the future.

COURSE SUBSTITUTIONS. Although the engineering course requirements may seem very rigid to you, most universities have a mechanism for substituting one course for another. For example, you may want to conduct an *independent study* with a professor rather than taking a specific required course. Or you may want to substitute a course taught by the economics department for the required course in engineering economics. Generally, such substitutions can be made if you gain necessary approvals.

OVERLOAD POLICY. Do you know the maximum number of units you can register for? If you want to exceed that number, your university usually has a procedure for you to seek approval to do so. And if your GPA is high, approval will probably be granted.

CREDIT FOR COURSES AT OTHER INSTITUTIONS. You may want to take a course at a community college or other four-year university during the summer. Before you do so, be sure to check out your university's policy on this. Most likely, you must get written approval in advance if you expect to receive transfer credit when you return.

STUDENT RIGHTS

There are regulations, policies, and procedures in the area of *student rights* that you should know about. Most universities have a *Statement of Students Rights*. For example, my university puts forth the following statement to students:

1. You have a right to **receive advisement** about your academic program, your career goals, and university policies and procedures.
2. In the classroom, you have the right to **express your views**, **receive instruction**, and **be graded fairly**.
3. You have a right to **form and participate in clubs and organizations** regardless of the interests those organizations promote.
4. You have the right to **publish or broadcast your opinions or concerns** to the campus community as long as they follow the rules of responsible journalism.

Check to see if your university or college has a similar statement of student rights. Make sure you know what your rights are!

Let's look briefly at student rights in three important categories:

(1) Petitions

(2) Grievances

(3) Privacy of Student Records

PETITIONS. In the previous sections, we gave examples of the many rules, regulations, policies, and procedures that are traditionally part of most educational systems. However, as the saying goes, "Every rule is made to be broken." If you find yourself constrained by a rule or regulation in a way that just doesn't make sense, you do have a recourse. Every university has a *petition for waiver of regulations* policy. My university catalog, for example, states:

> "Students who believe that extenuating circumstances might justify the waiver of a particular regulation or requirement may file a petition at their major department according to established procedures, for consideration by a faculty committee."

The particular approval process can vary from case to case. Suffice it to say, if the necessary signatures can be obtained, anything is possible.

STUDENT GRIEVANCES. Grievances are formal complaints by students against the university. The complaint might be about a specific instructor or administrator. Grievances generally involve an allegation by a student of unauthorized or unjustified actions that adversely affect the student's status, rights, or privileges, including but not limited to actions based on race, color, religion, sex, sexual orientation, national origin, age, disability, or veteran status.

You should check on your university's student grievance policy. Generally, such policies outline specific processes for filing student grievances. At my university, for example, the student grievance process has five steps. First, the student is required to attempt to resolve the grievance informally with the faculty member. If this is not satisfactory, the student is then required to seek the help of the department chair to resolve the grievance informally. If that doesn't solve the problem, the student must file a formal written grievance with the department chair, who may appoint a committee to make a recommendation. The fourth step, if necessary, is to contact the school dean, who will seek the recommendation of a school-wide committee The fifth and last step is to notify the university student grievance committee—only if the previous four steps have failed to resolve the issue.

PRIVACY OF STUDENT RECORDS. Your university maintains various types of records about you, such as academic records, financial aid records, health center records, and employment records. The Family Educational Rights and Privacy Act (FERPA) is a law designed to protect the privacy of your educational records. Following are your legal rights regarding these records:

(1) You have the right to inspect and review all of your educational records maintained by the university.

(2) You have the right to request that the university correct records you believe to be inaccurate or misleading.

(3) Generally, the university must have written permission from you before releasing any information from your records. (Note: The law does permit the university to disclose your records, without your consent, to certain parties under certain circumstances.)

8.6 STUDENT CONDUCT AND ETHICS

Along with rights come responsibilities. Your university has a code of conduct that delineates actions on your part that can result in disciplinary action. As an example, the following is a list of actions that warrant disciplinary action for all public universities in California [8]:

- Cheating, plagiarism, or other forms of academic dishonesty that are intended to gain unfair academic advantage
- Furnishing false information to a campus official, faculty member, or campus office
- Forgery, alteration, or misuse of a campus document, key, or identification instrument
- Misrepresentation of one's self or an organization to be an authorized agent of a campus
- Obstruction or disruption, on or off campus property, of the campus educational process, administrative process, or other campus function
- Theft of, or non-accidental damage to, campus property, or property in the possession of, or owned by, a member of the campus community
- Unauthorized entry into, presence in, use of, or misuse of campus property
- Use, possession, manufacture, or distribution of illegal drugs or drug-related paraphernalia, or the misuse of legal pharmaceutical drugs
- Use, possession, manufacture, or distribution of alcoholic beverages, or public intoxication while on campus or at a campus-related activity
- Possession or misuse of firearms or guns, replicas, ammunition, explosives, fireworks, knives, other weapons, or dangerous chemicals on campus or at a campus-related activity
- Disorderly, lewd, indecent, or obscene behavior at a campus-related activity, or directed toward a member of the campus community
- Conduct that threatens or endangers the health or safety of any person within or related to the campus community, including physical abuse, threats, intimidation, harassment, or sexual misconduct
- Misuse of campus computer facilities or resources including: use of another's identification or password; unauthorized entry into or transfer of a file; violation of copyright laws; sending obscene,

intimidating, or abusive messages; interference with the work of another member of the campus community or with campus operations

- Violation of any published campus policy, rule, regulation, or presidential order
- Encouraging, permitting, or assisting another to do any act that could subject him or her to discipline

Each of the above behaviors can bring about disciplinary sanctions including assignment of a failing grade in a course, probation, suspension, or expulsion. Many of these acts are also crimes that can result in criminal prosecution in addition to university discipline.

Of these actions, the one that occurs most often is academic dishonesty. Because of its importance, let's review what this entails.

ACADEMIC DISHONESTY

During your engineering study, you will address the topic of engineering ethics. *Ethics* is the study of what, on a social level, is right and wrong. Engineering ethics considers how engineers should behave in different situations—what behaviors are right and what are wrong.

The preamble to the Code of Ethics of the National Society of Professional Engineers (NSPE) provides a perspective on the importance of ethics in engineering:

> "Engineering is an important and learned profession. As members of this profession, engineers are expected to exhibit the highest standards of honesty and integrity. Engineering has a direct and vital impact on the quality of life for all people. Accordingly, the services provided by engineers require honesty, impartiality, fairness, and equity, and must be dedicated to the protection of the public health, safety, and welfare. Engineers must perform under a standard of professional behavior that requires adherence to the highest principles of ethical conduct."

I would encourage you to review the NSPE *Code of Ethics for Engineers* at: http://nspe.org/ethics/eh1-code.asp. The Code will give you an idea of the high standard of behavior expected of engineers. Hopefully, doing so will motivate you to begin to practice high standards of honesty and integrity during your tenure as an engineering student.

Ethics is a difficult subject because it is not always clear whether a certain behavior is ethical or unethical. Often engineers face *dilemmas*—problems for which there are no satisfactory solutions. Thus we are faced with making the "least" unethical choice. As a student, you will also face ethical dilemmas. Consider the following examples:

- You inadvertently saw several of the problems on an upcoming exam when you visited your professor in her office.
- Your professor incorrectly totaled the points on your midterm, giving you a 78 when you really only scored 58.
- A friend has been sick and asks to copy your homework that is due in a few hours.
- You have a lousy professor who gives you a student opinion survey at the end of the course to evaluate his teaching. He asks you to complete it and insert it in your final exam.
- Your professor has announced that her office hours are MW 10 a.m. - 12 noon. You have gone to her office during this time interval on four occasions and she has not been there.
- The data from your laboratory experiment doesn't make any sense. Your lab partner brings you a lab report from last term and suggests that you just use the data from that report.
- Your dean invites you to be part of a group of students to meet with the chair of the ABET visiting team. The dean asks you not to say anything negative about the engineering program.
- You notice two students in your class exchange their test papers during the final exam.
- You work part-time as a student assistant in the department office. The department secretary tells you to feel free to take whatever office supplies (paper, pencils, etc) you need from the department office supply stock room.

REFLECTION

Reflect on the ethical dilemmas just presented. In each of these situations, do you see that there are alternate courses of action? Do you see that the alternate courses of action reflect competing values (e.g., pleasing someone else vs. doing the right thing)? What would you do in each of these situations? Note that in some cases, it is very easy to decide what's right. In other cases, it is much more difficult.

There are, however, some areas of academic honesty for which there is no confusion over right and wrong. These include cheating, fabrication, facilitating academic dishonesty, and plagiarism.

CHEATING. *Cheating* is intentionally using or attempting to use unauthorized materials, information, or study aids in any academic exercise. Specific examples of cheating are:

- Receiving or knowingly supplying unauthorized information during an examination
- Using unauthorized material/sources during an examination
- Changing an answer after work has been graded, and presenting it as improperly graded
- Taking an examination for another student or having another student take an examination for you
- Feigning illness or telling falsehoods to avoid taking an examination at the scheduled time
- Forging or altering registration or grade documents

FABRICATION. *Fabrication* is the intentional, unauthorized falsification or invention of information or citations in an academic exercise. One example would be to make up or alter laboratory data.

FACILITATING ACADEMIC DISHONESTY. *Facilitating academic dishonesty* is intentionally or knowingly helping or attempting to help another to commit an act of academic dishonesty.

PLAGIARISM. *Plagiarism* is intentionally or knowingly representing the works or ideas of another as one's own in any academic exercise. The most extreme forms of plagiarism are the use of a paper written by another person or obtained from a commercial source, or the use of passages copied word for word without attributing the passage to the writer.

8.7 Graduate Study in Engineering

Most of our discussion to this point has been directed at completing your B.S. degree in engineering. When you do complete your undergraduate work, a variety of options await you. You can go to work as a practicing engineer in industry or government, or you can continue your education by working toward a graduate degree. The graduate degree could be in engineering or in other areas such as business, law, or medicine. Several of these opportunities for continuing your education are discussed in the following sections.

Benefits of Graduate Study in Engineering

Continuing your study through an M.S. or Ph.D. degree in engineering is an invaluable investment in yourself and in your future, regardless of what you plan to do professionally. The additional years you devote to graduate study will pay off again and again throughout your career in the following ways:

- You will further bolster your self-esteem and self-confidence.
- You will broaden your career choices and open doors to more challenging jobs—either in academe or in industry.
- You will increase your potential earnings over your lifetime.
- You will gain increased prestige, and others will accord you more respect.

I can't tell you how many times I have seen people's opinion of me rise when they learn that I have a Ph.D. in engineering. I can say without reservation that I would have missed much of what has been significant in my life had I not decided to seek a Ph.D.

M.S. Degree in Engineering

The first degree after the B.S. degree is the Master of Science Degree (M.S.). Getting your master's degree takes about one year of full-time study or two years of part-time study. There are three possible options for obtaining an M.S. degree: (1) all coursework; (2) coursework plus a project; or (3) coursework plus a thesis. Some engineering colleges may offer only one option, while others offer two of the three options or even all three.

Generally, option #1 requires about ten courses; option #2 about nine courses plus a project; and option #3 about eight courses plus a thesis. A limited number of these courses can be at the senior level with the remainder at the graduate level.

The project and thesis are similar but differ in both the type and amount of work. The project can generally be completed in one term and tends to be more practical. The thesis generally takes two or more terms and involves the development of new knowledge through research.

PH.D. DEGREE IN ENGINEERING

The Doctor of Philosophy (Ph.D.) degree is the highest educational degree in engineering. It generally takes four to five years or longer of full-time study beyond the B.S. degree. Typically, a Ph.D. program consists of about two years of coursework, culminating in comprehensive examinations (*comps*) covering your areas of specialty. After you pass the *comps*, you work full-time on a major research project, which becomes your Ph.D. dissertation.

Normally, you would apply for admission to a Ph.D. program following completion of your M.S. degree. If you continue on at the same institution, you will generally save time. You can also complete the M.S. degree at one university and then move to another for the Ph.D. While this can be a broadening experience, it will generally extend the time required to obtain the Ph.D. degree.

At some institutions, you can be admitted directly into the Ph.D. program upon completion of your B.S. degree. In some cases you simply "pick up" the M.S. along the way with little or no additional work. At others places you may be required to take a special exam or complete a thesis to get the M.S. degree. At still other places, you can elect to skip the M.S. degree completely.

The Ph.D. degree can prepare you for a career either in industry or academe. A career as an engineering professor can provide special rewards. If you would like to know more about these rewards, I suggest that you read "An Academic Career: It Could Be for You" [9].

FULL-TIME OR PART-TIME?

It is possible to work full-time in industry and pursue graduate study in engineering on a part-time basis. This is the way many engineers obtain their M.S. degree. But it is much more difficult, if not impossible, to complete a Ph.D. degree on a part-time basis. Whether earning an M.S. or a Ph.D., I certainly advise you to consider full-time graduate study if you can arrange it. Many of the benefits of graduate education come from being fully immersed in the academic environment—from concentrated study in your area of specialty; from engaging in dialogue and working closely with faculty and other graduate students; and from carrying out research under the close supervision of a faculty advisor.

There may be some benefits to working full-time in industry for a period after you receive your B.S. degree and then returning to full-time graduate study. You may want a break from school. You may have incurred debts that you need to pay off. You may be anxious to apply what you have learned. There is, however, a potential problem with working too long in industry. You may get used to a full-time engineering salary, making it difficult to return to the more modest student life. Of course, this depends on you and your commitment to your education.

How Will You Support Yourself?

Graduate study is different from undergraduate study in that there is a good chance you will be paid to do it. Any engineering graduate who has the potential to get a Ph.D. has a good chance of lining up adequate financial support for full-time graduate study. There are three kinds of financial support for graduate study: fellowships, teaching assistantships, and research assistantships. All three usually cover tuition and fees and provide a stipend for living expenses. Although fellowships and assistantships provide you much less money than full-time industrial positions, they usually support you adequately to work full-time on your degree.

8.8 Engineering Study as Preparation for Other Careers

In Chapter 2, we made the point that an undergraduate engineering education is excellent preparation for whatever you want to do. Engineering study is particularly good preparation for graduate study in related fields such as business, law, and medicine. Each of these opportunities is discussed below.

Master of Business Administration (MBA)

One of the engineering job functions described in Chapter 2 was management, which typically involves either *line supervision* or *project management*. Your engineering education will not prepare you fully for these management functions. As a manager you may very well need background in economics, accounting, finance, marketing, business law, and personnel management. You will receive very little, if any, training in these subjects as part of your engineering program. The ideal academic program to give you this additional background is the Master of Business Administration (MBA).

The MBA differs from the M.S. degree in business administration. Whereas the M.S. degree in business is designed for those who did their undergraduate work in business administration, the MBA is designed for those who did their undergraduate work in other academic fields.

Admission to an MBA program does not require any prior background in business administration. Completing the MBA takes two years of full-time study. The first year is spent developing your background in accounting, economics, marketing, business law, finance, computer information systems, and management. The second year is devoted to more advanced study in these subjects, with the opportunity to specialize in one of them.

Admission to an MBA program is based on your undergraduate record in engineering, letters of recommendation, and scores on the Graduate Management Admission Test (GMAT). The GMAT is a national standardized test administered by the Graduation Management Admission Council (GMAC) and is offered year-round at test centers throughout the world. The test covers quantitative, verbal, and analytical writing skills.

You can learn about MBA programs and the GMAT exam by visiting the Graduatc Management Admission Council (GMAC) web site:

http://www.mba.com

Engineering students generally do very well both on the GMAT and in MBA programs. The mathematical background and strong problem-solving skills gained through an undergraduate engineering education are excellent preparation for the MBA program. If you wish to prepare further for an MBA, you should use any free electives you have as an undergraduate to take courses in economics, accounting, or behavioral science.

There are two schools of thought regarding the best path to an MBA degree. One is that you should first work for several years in a technical position to gain professional engineering experience. If you are chosen for management or decide you want to seek a management position, you would then pursue the MBA either part-time while continuing to work or by returning to school full-time.

The second school of thought is that it is better to get the MBA prior to entering the workforce. The combination of an engineering degree and MBA could lead you directly into an entry-level management position, but if it doesn't, you will be in a good position to land a management position within a short time.

One last thing regarding the MBA. If you have any thoughts of eventually starting your own company, the MBA will provide you excellent training for doing so. Being a successful entrepreneur requires competence in finance, accounting, marketing, business law, and personnel management—all areas that receive significant coverage in the MBA program.

LAW

Excellent opportunities exist for engineers in the law profession. The primary opportunity is in patent law, where technical expertise combined with legal knowledge are essential. But other legal specialties such as environmental law and product liability law also fit well with an engineering background.

There are no specific undergraduate course requirements for law school. The most traditional pre-law majors include history, English, philosophy, political science, economics, and business, but any major including engineering is acceptable.

According to the American Bar Association (ABA) [10]:

> "The core skills and values that are essential for competent lawyering include analytic and problem-solving skills, critical reading abilities, writing skills, oral communication and listening abilities, general research skills, [and] task organization and management skills . . . . "

This would suggest that engineering study is, indeed, excellent preparation for law school, since an engineering graduate would already possess many of these skills.

Admission to one of the 194 ABA-approved law schools in the U.S. is based on undergraduate transcripts, letters of recommendation, and scores on the Law School Admissions Test (LSAT). The LSAT is a multiple-choice exam designed to measure the following skills: reading and comprehension of complex texts with accuracy and insight; organization and management of information and the ability to draw reasonable inferences from it; the ability to reason critically; and the analysis and evaluation of the reasoning and arguments of others. The test is administered by the Law School Admission Council (LSAC) and is offered four times a year at designated centers throughout the world.

You can obtain information about law schools and the LSAT at the Law School Admission Council web site:

http://www.lsac.org

If you are interested in law school, you should concentrate your elective undergraduate courses in history, economics, political science, and logic. Strong reading, writing, and oral communication skills are also important. Any opportunities you have to gain familiarity with legal terminology and the judicial process will also be beneficial in law school.

MEDICINE

Engineering study is also excellent preparation for medical school. Of the 17,003 students who entered the nation's 125 medical schools in 2005, 983 (5.8 percent) received their undergraduate degree in engineering [11]. Surprisingly, engineering ranked third (behind biology and biochemistry) in undergraduate majors for the 2005 medical school class. Perhaps more significant is that 54 percent of engineering majors who applied were accepted and entered medical school, compared to 45 percent of all other majors.

It is no accident that medical school admissions rates are high for engineering graduates. The logical thinking and problem-solving skills developed through engineering study have a direct carry-over to the diagnostic skills practiced by physicians. The combination of engineering and medicine can lead to careers in medical research or in the development of biomedical devices and equipment.

Engineering has a particular benefit over the more traditional pre-med majors such as biology, chemistry, and health science. Engineering offers students an excellent "fall-back" career option if they are either unable to gain admission to medical school or lose interest in a medical career.

Medical school admission requirements vary from school to school. Undergraduate course requirements include one year of biology, one year of physics, one year of English, and two years of chemistry. Experience in the health professions, extracurricular activities, and work experience are also encouraged.

Admission to medical school is based on undergraduate grades, scores on the Medical College Admission Test (MCAT), letters of recommendation, a personal statement, and a personal interview. The MCAT contains four sections: (1) verbal reasoning; (2) physical sciences; (3) writing sample; and (4) biological sciences. In 2006-07, the exam will transition from a paper and pencil exam offered twice yearly to a computer-based exam offered multiple times each year at testing sites throughout the world.

You can get information about medical schools, the MCAT exam, and the application process by visiting the Association of American Medical Colleges (AAMC) web site:

http://www.aamc.org

The calculus, chemistry, and physics required in the engineering curriculum provide much of what is needed to prepare for the MCAT and to be admitted to medical school. Additional requirements in biology and chemistry must be taken either as elective courses or as extra courses. Biomedical engineering and chemical engineering are the two engineering disciplines that best meet the needs of a pre-med program, because additional biology and chemistry courses are part of the required curriculum.

REFLECTION

Reflect on the opportunities presented in the previous section for continuing your education beyond your B.S. degree in engineering. Does any one of them appeal to you? Getting your M.S. or Ph.D. degree in engineering? Getting your MBA? Going to law school? Going to medical school? If one does, what do you need to do in the next few years to ensure that you will be prepared for and will be admitted to post-graduate study in that field?

SUMMARY

The purpose of this chapter was to orient you to the engineering education system. By understanding that system, you will be better able to make it work for you.

First, we described how the engineering college (or school) fits into the overall organization of the university. We then discussed the role of community colleges in delivering the first two years of the engineering curriculum.

Next, we reviewed the criteria that each engineering program must meet or exceed to receive accreditation from the Accreditation Board for

Engineering and Technology (ABET). The various criteria apply to eight areas: students, program educational objectives, program outcomes and assessment, professional component, faculty, facilities, institutional support and resources, and program criteria.

We also discussed academic advising, which addresses curricular and career matters. Ways were outlined for you to ensure that you receive sound academic advising, regardless of the advising system in place at your engineering college.

Then we described various academic regulations, policies, and procedures. By understanding what these entail, you can ensure that the system works for you, not against you. We also discussed the important area of student rights, including the right to petition, the right to file grievances, and the right to privacy of records.

Along with these rights comes responsibility. We therefore discussed the responsibility of students to behave ethically and honestly.

Finally, we discussed opportunities to continue your education beyond the B.S. degree. Graduate study in engineering can lead you to M.S. and Ph.D. degrees. Opportunities to seek post-graduate education in other professional fields including business administration, law, and medicine were also described.

REFERENCES

1. *Digest of Education Statistics Tables and Figures, 2005,* National Center for Education Statistics, U.S. Department of Education, August, 2006. (http://nces.ed.gov/programs/digest/d05_tf.asp)
2. *ABET Accreditation Yearbook (2005),* Accreditation Board for Engineering and Technology, 111 Market Place, Suite 1050, Baltimore, MD, 2005. (http://www.abet.org/accrediteac.asp)
3. Tsapogas, John, "The Role of Community Colleges in the Education of Recent Science and Engineering Graduates," NSF 04-315, National Science Foundation, Washington, D.C., May 2004.
4. Wolf, Lawrence J., "The Added Value of Engineering Technology," Oregon Institute of Technology, Klamath Falls, OR.
5. Cheshier, Stephen R., *Studying Engineering Technology: A Blueprint for Success*, Discovery Press, Los Angeles, CA, 1998.
6. "Criteria for Accrediting Engineering Programs: Effective for Evaluations During the 2006-2007 Accreditation Cycle,"

Accreditation Board for Engineering and Technology, Baltimore, MD. (Available at: http://www.abet.org/forms.shtml)

7. Wankat, P. and Oreovicz, F., *Teaching Engineering*, McGraw-Hill, New York, NY, 1997.
8. "Title 5 – California Code of Regulations, Section 41301," California Department of Education, Sacramento, CA. (See: http://www.calstatela.edu/univ/stuaffrs/jao/doc/sfsc.pdf)
9. Landis, Raymond B., "An Academic Career: It Could Be for You," *Engineering Education,* July/August 1989. (Available from American Society for Engineering Education, Washington, D.C.)
10. *ABA - LSAC Official Guide to ABA-Approved Law Schools 2007,* Law School Admission Council, April 2006. (Available on Amazon.com)
11. "AAMC Data Warehouse: Applicant Matriculant File," Association of American Medical Colleges. Personal communication with Collins Mikesell (http://www.aamc.org/data/facts/start.htm)

PROBLEMS

1. Find out the names of the people in the following positions:
   a. The chair or head of your engineering department
   b. The dean of your engineering school or college
   c. The vice chancellor or vice president for academic affairs
   d. The chancellor or president of your university
2. Locate and read the following:
   a. The mission statement of your institution
   b. Educational objectives and program outcomes of your engineering program

   Are the educational objectives and program outcomes consistent with the mission statement of the institution? Write a one-page paper discussing how they are or are not consistent.
3. Rank the (a) - (k) outcomes in Criterion 3 of the ABET Engineering Criteria 2000 in order of importance in preparing an engineering graduate for a productive career as a practicing engineer. Prepare a two-minute talk describing why you ranked the #1 item as most important.

4. Visit two engineering professors during their office hours. Ask each to rank the (a) - (k) outcomes in Criterion 3 of the ABET Engineering Criteria 2000 in order of importance. Ask each to explain why they ranked their #1 item as most important. How does their ranking compare to yours? Prepare a five-minute presentation about what you learned from this exercise.

5. Identify the courses in your engineering curriculum that meet the ABET requirement for one year of mathematics and basic sciences. Do these courses equal or exceed 32 semester units (or 48 quarter units)?

6. Find out where you will learn the following computer skills in your engineering curriculum:

   a. Programming languages
   b. Word processing
   c. Computer-aided design
   d. Spread sheets
   e. Database management systems
   f. Computer graphics
   g. Data acquisition

7. Assuming you have already devised a "road map" showing the courses you will take to meet the requirements for your engineering major, review this plan with your academic advisor and revise it based on the feedback you receive.

8. Research the academic advising system in place in your engineering college. Write a one-page description of that advising system, including your critique of how well it works for students.

9. Does either your engineering college or your engineering department publish an *Engineering Student Handbook*? Obtain a copy and read it thoroughly. Write a one-page paper summarizing what you learned that will be of benefit to you.

10. Find out your university's regulations regarding the following academic issues:

    a. Taking courses *Credit/No Credit*
    b. Incompletes
    c. Repeat Grade Policy
    d. Credit by Examination
    e. Probation

f. Disqualification
g. Dean's List
h. Honors at graduation

11. After you have completed 100 units, your overall GPA is 2.4. During the next term you take 16 units and achieve a 3.4 GPA for the term. What is your overall GPA then? If your overall GPA was 3.4 after 100 units and you take 16 units and make a 2.4 GPA for the term, what is your overall GPA then? What is the point or *message* of this exercise?

12. Determine whether your university has a S*tatement of Student Rights*. Obtain a copy and compare it to the rights discussed in Section 8.5.

13. Determine whether your university has a S*tudent Code of Conduct*. If it does, obtain a copy and review the list of actions that warrant disciplinary action.

14. Write a brief opinion as to how you would handle each of the ethical dilemmas posed in Section 8.6. Discuss your responses with at least one other student.

15. Find out how you go about changing your major from one engineering discipline to another at your institution. Are some disciplines more difficult to get into than others? Which discipline is the most difficult to get into?

16. Investigate the graduate programs available in engineering at your institution. Which engineering programs offer M.S. degrees? Which engineering programs offer Ph.D. degrees? Write down the requirements (e.g., GPA, Graduate Record Examination scores, etc.) one needs to be admitted to a graduate program in your major.

17. Consider whether you are interested in pursuing one of the three non-engineering careers discussed in Section 8.8. Locate an advisor in the area of your greatest interest (MBA, pre-law, pre-medicine) and seek additional information.

APPENDIX A

21 Definitions of Engineering

The following 21 definitions of engineering from such notable sources as Count Rumford to Samuel C. Florman were complied by Harry T. Roman of East Orange, N.J., USA.

The application of science to the common purpose of life.

--Count Rumford (1799)

Engineering is the art of directing the great sources of power in nature for the use and convenience of man.

--Thomas Tredgold (1828)

It would be well if engineering were less generally thought of, and even defined, as the art of constructing. In a certain sense it is rather the art of not constructing; or, to define it rudely but not inaptly, it is the art of doing that well with one dollar which any bungler can do with two after a fashion.

--A. M. Wellington (1887)

Engineering is the art of organizing and directing men and controlling the forces and materials of nature for the benefit of the human race.

--Henry G. Stott (1907)

Engineering is the science of economy, of conserving the energy, kinetic and potential, provided and stored up by nature for the use of man. It is the business of engineering to utilize this energy to the best advantage, so that there may be the least possible waste.

--Willard A. Smith (1908)

Engineering is the conscious application of science to the problems of economic production.

--H. P. Gillette (1910)

Engineering is the art or science of utilizing, directing or instructing others in the utilization of the principles, forces, properties and substance of nature in the production, manufacture, construction, operation and use of things . . . or of means, methods, machines, devices and structures.

--Alfred W. Kiddle (1920)

Engineering is the practice of safe and economic application of the scientific laws governing the forces and materials of nature by means of organization, design and construction, for the general benefit of mankind.

--S. E. Lindsay (1920)

Engineering is an activity other than purely manual and physical work which brings about the utilization of the materials and laws of nature for the good of humanity.

--R. E. Hellmund (1929)

Engineering is the science and art of efficient dealing with materials and forces . . . it involves the most economic design and execution . . . assuring, when properly performed, the most advantageous combination of accuracy, safety, durability, speed, simplicity, efficiency, and economy possible for the conditions of design and service.

--J. A. L. Waddell, Frank W. Skinner, and H. E. Wessman (1933)

Engineering is the professional and systematic application of science to the efficient utilization of natural resources to produce wealth.

--T. J. Hoover and J. C. L. Fish (1941)

The activity characteristic of professional engineering is the design of structures, machines, circuits, or processes, or of combinations of these elements into systems or plants and the analysis and prediction of their performance and costs under specified working conditions.

--M. P. O'Brien (1954)

The ideal engineer is a composite . . . He is not a scientist, he is not a mathematician, he is not a sociologist or a writer; but he may use the knowledge and techniques of any or all of these disciplines in solving engineering problems.

--N. W. Dougherty (1955)

Engineers participate in the activities which make the resources of nature available in a form beneficial to man and provide systems which will perform optimally and economically.

--L. M. K. Boelter (1957)

The engineer is the key figure in the material progress of the world. It is his engineering that makes a reality of the potential value of science by translating scientific knowledge into tools, resources, energy and labor to bring them into the service of man . . . To make contributions of this kind the engineer requires the imagination to visualize the needs of society and to appreciate what is possible as well as the technological and broad social age understanding to bring his vision to reality.

--Sir Eric Ashby (1958)

The engineer has been, and is, a maker of history.

--James Kip Finch (1960)

Engineering is the profession in which a knowledge of the mathematical and natural sciences gained by study, experience, and practice is applied with judgment to develop ways to utilize, economically, the materials and forces of nature for the benefit of mankind.

--Engineers Council for Professional Development (1961/1979)

Engineering is the professional art of applying science to the optimum conversion of natural resources to the benefit of man.

--Ralph J. Smith (1962)

Engineering is not merely knowing and being knowledgeable, like a walking encyclopedia; engineering is not merely analysis; engineering is not merely the possession of the capacity to get elegant solutions to non-existent engineering problems; engineering is practicing the art of the organized forcing of technological change . . . Engineers operate at the interface between science and society . . .

--Dean Gordon Brown; Massachusetts Institute of Technology (1962)

The story of civilization is, in a sense, the story of engineering - that long and arduous struggle to make the forces of nature work for man's good.

--L. Sprague DeCamp (1963)

Engineering is the art or science of making practical.

--Samuel C. Florman (1976)

APPENDIX B

Greatest Engineering Achievements of the 20th Century

#20 - HIGH PERFORMANCE MATERIALS

From the building blocks of iron and steel to the latest advances in polymers, ceramics, and composites, the 20th century has seen a revolution in materials. Engineers have tailored and enhanced material properties for uses in thousands of applications.

#19 - NUCLEAR TECHNOLOGIES

The harnessing of the atom changed the nature of war forever and astounded the world with its awesome power. Nuclear technologies also gave us a new source of electric power and new capabilities in medical research and imaging.

#18 - LASER AND FIBER OPTICS

Pulses of light from lasers are used in industrial tools, surgical devices, satellites, and other products. In communications, highly pure glass fibers now provide the infrastructure to carry information via laser-produced light—a revolutionary technical achievement. Today, a single fiber-optic cable can transmit tens of millions of phone calls, data files, and video images.

#17 - PETROLEUM AND GAS TECHNOLOGIES

Petroleum has been a critical component of 20th century life, providing fuel for cars, homes, and industries. Petrochemicals are used in products ranging from aspirin to zippers. Spurred on by engineering advances in oil exploration and processing, petroleum products have had an enormous impact on world economies, people, and politics.

#16 - HEALTH TECHNOLOGIES

Advances in 20th century medical technology have been astounding. Medical professionals now have an arsenal of diagnostic and treatment equipment at their disposal. Artificial organs, replacement joints, imaging technologies, and bio-materials are but a few of the engineered products that improve the quality of life for millions.

#15 - Household Appliances

Engineering innovation produced a wide variety of devices, including electric ranges, vacuum cleaners, dishwashers, and dryers. These and other products give us more free time, enable more people to work outside the home, and contribute significantly to our economy.

#14 - Imaging Technologies

From tiny atoms to distant galaxies, imaging technologies have expanded the reach of our vision. Probing the human body, mapping ocean floors, tracking weather patterns—all are the result of engineering advances in imaging technologies.

#13 - Internet

The Internet is changing business practices, educational pursuits, and personal communications. By providing global access to news, commerce, and vast stores of information, the Internet brings people together globally while adding convenience and efficiency to our lives.

#12 - Space Exploration

From early test rockets to sophisticated satellites, the human expansion into space is perhaps the most amazing engineering feat of the 20th century. The development of spacecraft has thrilled the world, expanded our knowledge base, and improved our capabilities. Thousands of useful products and services have resulted from the space program, including medical devices, improved weather forecasting, and wireless communications.

#11 - Interstate Highways

Highways provide one of our most cherished assets—the freedom of personal mobility. Thousands of engineers built the roads, bridges, and tunnels that connect our communities, enable goods and services to reach remote areas, encourage growth, and facilitate commerce.

#10 - Air Conditioning and Refrigeration

Air conditioning and refrigeration changed life immensely in the 20th century. Dozens of engineering innovations made it possible to transport and store fresh foods, for people to live and work comfortably in sweltering climates, and to create stable environments for the sensitive components that underlie today's information-technology economy.

#9 - TELEPHONE

The telephone is a cornerstone of modern life. Nearly instant connections—between friends, families, businesses, and nations—enable communications that enhance our lives, industries, and economies. With remarkable innovations, engineers have brought us from copper wire to fiber optics, from switchboards to satellites, and from party lines to the Internet.

#8 - COMPUTERS

The computer has transformed businesses and lives around the world by increasing productivity and opening access to vast amounts of knowledge. Computers have relieved the drudgery of routine daily tasks, and brought new ways to handle complex ones. Engineering ingenuity fueled this revolution, and continues to make computers faster, more powerful, and more affordable.

#7 - AGRICULTURAL MECHANIZATION

The machinery of farms—tractors, cultivators, combines, and hundreds of others—dramatically increased farm efficiency and productivity in the 20th century. At the start of the century, four U.S. farmers could feed about ten people. By the end, with the help of engineering innovation, a single farmer could feed more than 100 people.

#6 - RADIO AND TELEVISION

Radio and television were major agents of social change in the 20th century, opening windows to other lives, to remote areas of the world, and to history in the making. From wireless telegraph to today's advanced satellite systems, engineers have developed remarkable technologies that inform and entertain millions every day.

#5 - ELECTRONICS

Electronics provide the basis for countless innovations—CD players, TVs, and computers, to name a few. From vacuum tubes to transistors, to integrated circuits, engineers have made electronics smaller, more powerful, and more efficient, paving the way for products that have improved the quality and convenience of modern life.

#4 - SAFE AND ABUNDANT WATER

The availability of safe and abundant water literally changed the way Americans lived and died during the last century. In the early 1900s, waterborne diseases like typhoid fever and cholera killed tens-of-thousands of people annually, and dysentery and diarrhea, the most common waterborne diseases, were the third largest cause of death. By the 1940s, however, water treatment and distribution systems devised by engineers had almost totally eliminated these diseases in American and other developed nations. They also brought water to vast tracts of land that would otherwise have been uninhabitable.

#3 - AIRPLANE

Modern air travel transports goods and people quickly around the globe, facilitating our personal, cultural, and commercial interaction. Engineering innovation—from the Wright brothers' airplane to today's supersonic jets—have made it all possible.

#2 - AUTOMOBILE

The automobile may be the ultimate symbol of personal freedom. It's also the world's major transporter of people and goods, and a strong source of economic growth and stability. From early Tin Lizzies to today's sleek sedans, the automobile is a showcase of 20th century engineering ingenuity, with countless innovations made in design, production, and safety.

#1 - ELECTRIFICATION

Electrification powers almost every pursuit and enterprise in modern society. It has literally lighted the world and impacted countless areas of daily life, including food production and processing, air conditioning and heating, refrigeration, entertainment, transportation, communication, health care, and computers. Thousands of engineers made it happen, with innovative work in fuel sources, power generating techniques, and transmission grids.

Appendix C

Engineers Among the World's 200 Wealthiest Individuals

Name	Position	Wealth	Major	Institution
Carlos Slim Helu	Mexican entrepreneur and businessman	$30 billion	Engineering	Universidad Nacional Autonoma de Mexico
Bernard Arnault	CEO Louis Vuitton Moët Hennessy	$21.5 billion	Engineering	École Polytechnique
Azim Premji	Chairman and CEO Wipro Technologies	$13.3 billion	Engineering	Stanford University
Sergey Brin	Co-Founder of Google	$12.9 billion	Computer Science	University of Maryland
Larry Page	Co-Founder of Google	$12.8 billion	Computer Engineering	University of Michigan
Charles Koch	Chairman and CEO Koch Industries	$12 billion	Engineering	MIT
David H. Koch	Executive VP Koch Industries	$12 billion	Chemical Engineering	MIT
Vagit Alekperov	President of LUKoil Oil Company	$11 billion	Engineering	Azerbaijan Institute of Oil and Chemistry
Vladimir Lisin	Owner Novolipetsk steel mill	$10.7 billion	Engineering	Siberian Metallurgical Institute
Pierre Omidyar	Founder/Chairman eBay	$10.4 billion	Computer Science	Tufts University
Viktor Vekselberg	Chairman of Tyumen Oil (TNK)	$9.7 billion	Engineering	Moscow State University of Railroad Engineering
Mukesh Ambani	Chairman Reliance Industries Ltd	$8.5 billion	Chemical Engineering	University of Mumbai
Serge Dassault	Honorary Chairman Dassault Aviation	$8.5 billion	Aeronautical Engineering	École Polytechnique
Alexei Mordashov	Chairman of the Board of RUSAL	$7.6 billion	Engineering	M.V. Lomonosov Moscow State University

Name	Position	Wealth	Major	Institution
Abdul Aziz Al Ghurair	Chief Executive of Mashreqbank	$6.9 billion	Industrial Engineering	California Polytechnic State University
Stefan Quandt	Owner Delton AG	$6.6 billion	Engineering	Technical University of Karlsruhe
Michael Bloomberg	Mayor of New York City	$5.1 billion	Electrical Engineering	Johns Hopkins University
Jeffrey Skoll	Former President of eBay	$5 billion	Electrical Engineering	University of Toronto
Eric Schmidt	Chairman and CEO of Google	$4.8 billion	Electrical Engineering	Princeton University
Jeff Bezos	Founder and CEO of Amazon.com	$4.3 billion	Comp Science/ Electrical Engr	Princeton University
Shiv Nadar	CEO of Hindustan Computers Ltd (HCL)	$4 billion	Engineering	PSG College of Technology

INDEX

E

F

G

H

I

J

L

M